Where Am I?

Where Am I?

WHY WE CAN FIND OUR WAY TO THE MOON
BUT GET LOST IN THE MALL

Colin Ellard

HARPER PERENNIAL

Published by Harper Perennial, an imprint of HarperCollins Publishers Ltd

First published in hardcover by HarperCollins Publishers Ltd: 2009
This Harper Perennial trade paperback edition: 2010

Grateful acknowledgment is made for permission to reprint material from the following:
We, The Navigators: The Ancient Art of Landfilling in the Pacific,
by David Lewis, © 1972, reprinted by permission of University of Hawaii Press;
"Arabian Sands" by Wilfred Thesiger, reproduced with permission of Curtis Brown Group Ltd,
London, on behalf of the Estate of Wilfred Thesiger, © 1959 by Wilfred Thesiger;
Figures 10 and 11, John Peponis et al., "Designing Space to Support
Knowledge Work," *Environment and Behavior, 39* (6), p. 26, © 2007
by SAGE Publications, reprinted by permission;
Figures 14 and 15 reprinted from *Progress in Planning, 67* (3),
Bill Hillier and Laura Vaughan, "The City as
One Thing," pp. 205–230, © 2007 by Elsevier, reprinted by permission.

HarperCollins books may be purchased for educational, business,
or sales promotional use through our Special Markets Department.

HarperCollins Publishers Ltd
2 Bloor Street East, 20th Floor
Toronto, Ontario, Canada
M4W 1A8

www.harpercollins.ca

Library and Archives Canada Cataloguing in Publication

Ellard, Colin, 1958–
Where am I? : why we can find our way to the moon but get
lost in the mall / Colin Ellard.

ISBN 978-1-55468-394-9

1. Geographical perception. 2. Spatial behavior. 3. Cognitive maps
(Psychology). 4. Navigation—Psychological aspects. I. Title.

BF469.E45 2010   153.7'52   C2010-900803-0

Printed in the United States
RRD 10 9 8 7 6 5 4 3 2 1

For Karen.
Without you, I am nowhere.

CONTENTS

PART I: WHY ANTS DON'T GET LOST AT THE MALL
HOW HUMANS AND ANIMALS NAVIGATE SPACE

PART II: MAKING YOUR WAY IN THE WORLD TODAY

HOW OUR MIND SHAPES THE PLACES WHERE WE WORK, LIVE, AND PLAY

INTRODUCTION
LOST AND FOUND

There is a ritual that all parents must endure from time to time, known as the weekend camping trip. This involves piling the car high with everything from cooking utensils and plastic tarps to a few spare pairs of Bob the Builder underwear and then setting off for a drive to a local park where, more often than not, visions of campfires under starlit skies are replaced by shivering huddles in soggy, windblown tents. It is a brilliant tribute to the human spirit that we always come home from such adventures with nothing but happy memories and eager anticipation for the next close encounter with the Land. It was on one such excursion that I came into intimate contact with my own fragile grip on physical space.

My wife, Karen, and I had blown off the idea of a simple visit to a friendly park containing clearly marked camping plots, complete with firepits and driveways, user-friendly toilets and nearby convenience stores. Instead, we chose to drive for most of a day with friends of ours and our children to one of the northernmost parts of Algonquin Park—a piece of protected land in the heart of

Ontario with an area that is not much smaller than Portugal. Most of the park, populated by moose, wolves, deer, and the occasional black bear, is accessible only on foot or by canoe. We wanted our children to experience true adventure, so, along with our friends, we ventured into the woods with the barest of supplies and one small canoe. Our destination was a canoe-in campsite on the edge of a small lake. The setting was stunning. Our site looked out on a tiny island with one tall, bare tree containing an osprey nest. We were able to sit on the shore and watch the majestic birds fly off in search of food for their chicks.

Soon after we arrived, the rain began. Determined to make the best of things, we spent as much time as we could hiking, canoeing, and exploring. Between adventures, we huddled under a small blue tarp and wrung out wet clothes; when the children weren't watching, the adults passed a small silver flask of liquid warmth back and forth. On the second day of our trip, we planned an ambitious journey to visit a scenic waterfall. It was too far for our youngest daughters, Jessica and Rebecca, to walk, so they rode in the canoe with our two friends, and Karen and I set off on foot with our oldest daughter, Sarah. We had warned the children beforehand that there was a remote chance we would see a bear, and that the best deterrent was to make plenty of noise as we walked, preferably by singing and clapping. Truthfully, though, we knew that bear sightings were so rare here that many park regulars went years without spotting a single specimen. As much as anything, our safety lecture was meant to heighten our children's excitement and enjoyment of the trip.

Sarah was a teenager at the time and was in no temper for either singing or clapping. As she skulked along the trail some distance ahead of the rest of us, doing her best to pretend not to be with us, I ran to catch up to her, and told her that if she was determined to lead the group, she would need to make some sort of noise. She responded by dropping to the back of the pack, leaving me in front. A minute later, as I sang the theme song from *The Flintstones,* I noticed a sapling swing first toward me and then away. I heard a loud rustle just off to my left, but couldn't see anything. Whatever it was must have run off. I called back, "I think I just startled a big animal! Maybe a deer!" As I turned back to the trail, my view was occluded by the flank of a very large bear, close enough to me that I could have reached out to touch it. It may have been sleeping near the trail and, despite my racket, I think that I might have startled it. In as calm a voice as I could muster, I told Karen and Sarah to back down the trail slowly. I told them not to turn their backs on the bear and not to run. I did much the same, even though by this time the bear had crossed the trail and disappeared into the thicket of woods.

Much later, back at the campsite, our nerves calmed and our children asleep, we spent much time emptying the silver flask and reflecting on our close call and on our vulnerability to whatever unexpected curves nature might throw our way. Stripped of the normal network of support that urban human beings use to get themselves through a day, stumbling along a marked trail in the woods, every slight misstep might spell disaster. If I had walked at a different pace, in a different direction, or without making such a din, my life could have been ended by a single angry swipe of a bear's paw. Our fate hinged on a crude park map and good luck. What were we doing out here with our kids?

In the morning, we broke camp. Remembering the four trips across the lake in our tiny canoe to move people and gear between trail and campsite, I pulled out a tattered trail map and argued that if some of us blazed a trail through the woods behind the campsite, we ought to come across a trail that skirted the lake. Karen and I offered to make the trip with a full pack of gear and some of the children in order to shorten the time before we were all sitting in a dry room and digging into a hot breakfast. The route was simple, and is laid out roughly in Figure 1. All we would need to do would be to walk in a straight line for about 100 meters and then, once we found the trail, make a right turn. It is never a good idea to leave a marked trail in wilderness. Our fragile understanding of where we are can collapse quickly, leaving us lost, disoriented, and in peril. In this case, though, the simple shortcut seemed like such a no-lose proposition that, given the rain, the wind, and the hungry bellies, it seemed a risk worth taking.

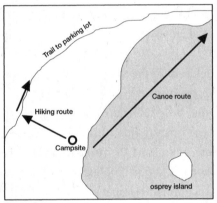

Figure 1: The ideal route to our campsite in Algonquin Park

About an hour after setting off on what should have been about a 45-minute hike, Karen and I knew something was wrong. We had crashed through thick underbrush, no easy feat while haul-ing a large backpack, and had found the trail exactly where we expected it. But as we walked along, the terrain looked less and less familiar. New lakes appeared where our map suggested none should exist. One of these lakes even had a small island with what looked much like a second osprey nest. It is a testament to the disorienting power of wilderness that this remarkable discovery did not give us an obvious clue as to where we were. Our feeling of unease was only increased when we encountered several piles of fresh bear droppings on the trail. After another fifteen minutes of walking, and with little awareness of our whereabouts, both Karen and I fought a rising tide of panic. We were lost. We were convinced that more bears were nearby. We had heavy gear, including some food that might draw the bears closer to us, and we had young children to protect. We stopped walking and did our best to put together a few rational thoughts. Eventually, we stumbled upon the mistake we had made. Figure 2 shows what we had done.

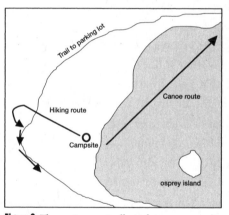

Figure 2: The route we actually took to our campsite in Algonquin Park

We had made critical errors: not only had we somehow crossed the trail without noticing that we had done so but we had then made a complete turnabout, all the while believing that we had been walking in a straight line. Anyone who has tried to walk in a straight line through a cluttered environment like a forest will not be surprised by this error. Even with extensive training, this is a difficult task. Try it. Close your eyes and walk in a straight line. Surprisingly difficult, isn't it? As if this wasn't bad enough, we had compounded our error by failing to recognize that the lake we saw with the second osprey island was the same lake that we had just spent two days camping beside. In fact, from the vantage point of the trail's edge where we stood trying to orient ourselves, we ought to have been able to *see* our old campsite. Eventually, after much head-scratching and a few more pauses to reassure ourselves that we now had our bearings, we stumbled into the parking lot at the end of the trail and drove, humbled, following an abundance of clearly marked road signs, to the local diner for a restoring feast.

What kind of creatures are we that we can design technology to navigate across oceans, continents, and even the forbidding reaches of space, yet we become hopelessly lost in a small forest? Why can't we walk in a straight line for more than a few steps or recognize landmarks that we've lived among for days?

You might say that what I've described is nothing more than the fumbling of a longtime urban dweller, off on a weekend lark with no training, no compass, and perhaps a bit too much good whiskey in his bloodstream to be able to keep his bearings. Yet both the lore and the science of human navigation suggest that compared to most other animals, human beings are mere beginners at navigation. In my laboratory at the University of Waterloo, the Research Laboratory for Immersive Virtual Environments (RELIVE), we have spent many years studying the performance of both animals and people in a great variety of navigation tasks. Some striking differences between species have emerged.

There is something of a paradox here. Our ability to *understand* physical space, the lengths and widths of objects, places, and even entire planets, is unrivaled among all of the animals. We make maps, draw graphs, and design sophisticated measuring machines. We have even launched arrays of satellites into outer space that we can use to guide the movements of everything from a jogger in the park to a jumbo jet or a gigantic cruise ship in the Mediterranean Sea.

As preponderantly cognitive beings, we bring our staggering intellect to bear on the technological conquest of space, yet we become lost in tiny patches of green space. We are forced to take frustrating interludes from our busy lives trying to locate our own car in a parking lot. We can even become lost in the spaces we build, such as office buildings, shopping malls, and hospitals. How do we reconcile these two basic facets of the human relationship with space and place—our theoretical mastery of abstract space and our ineffable clumsiness in finding our way? Is it possible that there might even be a connection between the two? Have we been so successful in building an environment that relieves us of the need to find our way from place to place that we have fundamentally changed our nature?

A black bear, like the one that I disturbed in the woods, can find its way home after being displaced distances of hundreds of kilometers using methods that have rebuffed the efforts of scientists to understand them.[1] What do bears know that we don't? How do monarch butterflies or migrating songbirds navigate even larger distances, thousands of kilometers in some instances, to targets that they might never even have seen before? How is it that a homing pigeon can be driven halfway across a continent in a light-proof box and then, on release, find its way unerringly back to its loft? How do newborn sea turtles waddle off a beach in Florida and migrate thousands of kilometers through ocean depths to rich foraging grounds near the coast of Africa? Perhaps most interestingly of all, how is it that the only existing animal that has come *close* to understanding how some of these magnificent navigational feats are performed is rendered helpless by a confusing thicket of woods or an unexpected hallway in an office building?

THINKING SPACE

Considered as a purely physical thing, space is the set of dimensions that defines locations, distances, and the relationship between one thing and another. Mathematicians and physicists tell us that there are many possible different kinds of spaces, each with their own set of dimensions and rules, but the space that operates on our biology is much like that envisioned by Euclid in his version of the geometry of space that we all learn in elementary school. Like all other animals, we live with the everyday reality that the dividing line between those things that are possible and those that are not is sharply drawn by the nature of space and time. I cannot be in both Chicago and Toronto at the same time, and my ability to move from one place to another, and the route that I must take to get there, depends on mathematical geometry and physical law.

There is, of course, much more to space than the pragmatics of getting from place to place. Since the dawn of thought, philosophers and scientists have struggled to define and to understand what space is. Our creation stories, whether they come from the ancient Greeks, the Indian Rig Veda, or the Old Testament, help us to come to terms with the idea that something came from nothing, and in managing this gigantic mental step, we have no choice but to ruminate on what the real difference is between nothing and something. Modern philosophers have had to struggle with such basic metaphysical questions about the nature of space in a context that is increasingly constrained by the discoveries of science.[2] Some physicists now tell us that spaces with twenty dimensions, in which parallel lines might curve and meet, may be much more than the fancies of mathematicians and philosophers, and may be physically real, whether we are able to perceive them or not. Classical ideas about time as the gatekeeper of events and the arbiter of simultaneity are also becoming frayed at the edges, as new discoveries in quantum physics suggest that there can be connections between objects separated by vast expanses of physical space, as if information can be passed from place to place through the universe with *no* intervening passage of time.[3]

Psychologists and other social scientists have long understood that there is much more to space than a set of points drawn on a graph or a map. Our language is inundated with spatial metaphors ("Let's not get ahead of ourselves," "The case is under review," "His behavior was over the top"), and our everyday understanding of space often departs from Euclidean order in spectacular fashion. Think of the way that your understanding of a new and unfamiliar space—a new home or place of work—changes over the course of your experience. The space itself does not change, but the mental representation of it is transformed dramatically.

Throughout much of this book, we will be preoccupied with questions related to navigation and wayfinding. Where am I? Where are *you* and how do you know? We will explore some of the details of the tools that both we and other animals use to find our way through spaces from one location to another. Though we will find much common ground between ourselves and other creatures of field and forest, some stark differences will emerge between our abilities and those of the many other animals with whom we share the planet. Some animals possess specialized senses and abilities that help them to know where they are. We do not. But we have come to an entirely new type of relationship with physical space. This new relationship, the markings of which are literally written into parts of our brain, not only allows us to cope with the problems of space but also in a way liberates us from space. It is as if our prodigious brains have grown to allow us to stand outside of space so that even as we struggle to find ways to survive in it, we can reflect on it at a distance, even reduce it to a set of mathematical formulas and abstract maps. It is this ability to stand outside real space mentally that has allowed us to conceive, develop, and use the technologies—rapid transit and communication, mass media, virtual reality—that have allowed us to further unshackle ourselves from space. In the chapters that follow, we will look at some of these space-bending technologies and the influence they have had on the way we live our lives, effects that extend from the way we design our homes and cities to the ways that we work, communicate, and play with one another.

In the first part of the book, we will survey the types of spatial information that are available to all animals as they move about the planet, and how they take advantage of them to find their way. We will begin by exploring the very simplest forms of navigation: how do we make our way from our current position to a target that we can see? At this early station point in our journey, we will see that we share much with all animals, from single-celled bacteria to lumbering bears and buzzing bees. In later chapters, we will explore more complicated navigational routines. Landmarks aren't always the targets of our movements, but they can be used to lead us to our targets ("Just keep the mountains on your left and you can't miss it"). Maps are endemic to navigators, but there are different kinds of maps. The maps we find in the glove compartments of our cars can be very different from those we hold in our minds as we try to give a set of directions. In turn, our maps can be different from those put together with a handful of neurons in the speck-sized brain of a bee. When we are trying to find our way through unfamiliar terrain, we may try to maintain our orientation by attending to and memorizing our outbound route ("Two lefts and a right at the next light"). We share this ability with many animals, but some animals, such as the desert ants of Northern Africa, can manage this kind of trip with exacting precision over staggering expanses of space, as if they possess tiny odometers that click off the miles with clockwork precision.

As we explore the details of these different means of wayfinding, we will gradually mark out the boundary that separates human abilities to deal with problems of space from those of other animals. Some animals rely on special senses, such as magnetic field sensitivity or the ability to analyze special properties of light waves that we cannot see. Other animals possess prodigious talents for memorizing their paths or the appearance of thousands of subtle landmarks in a seemingly featureless expanse of forest or meadow. Human beings follow a somewhat different course. Though some of us, particularly ancient wayfinders in preliterate societies, have been able to train ourselves to be sensitive to subtle perceptual cues about locations and routes, we more commonly find our way by connecting different types of images of places with stories that link one image to another. This kind of navigation can sometimes get us effectively from one place to another, but we sacrifice much of our understanding of how places are connected, the geometry of the world, in favor of a simpler view of the spaces we inhabit as a series of connected nodes. We can learn *what* is connected, but our command of the hows and wheres of such connections is shaky at best.

In the second part of the book, we will explore the implications of the differences between human beings and other animals for problems that involve our understanding of where we are. How do the sizes and shapes of our dwellings, offices, factories, civic buildings, and cities reflect our abilities (or inabilities) to come to terms with physical space? How does our unique conception of space as a topology of connected nodes influence the way that we interact with colleagues at work? How has modern technology, especially instantaneous communication using everything from the telephone to the Internet, changed our understanding and use of physical space? Has the fact that our minds are fuzzy about the geometry of the everyday world accelerated the extent to which such technologies have penetrated our lives? Would an animal that understood in its bones the meaning of space have been able to adapt to the hyperspace of the World Wide Web and virtual environments spanning the globe?

RECAPTURING PLACE BY RETHINKING SPACE

There can be little doubt that our ability to stand outside of real space and look back into it, reflect upon it, and shape it to our own designs and purposes has been responsible for much of the form of modern life. We have used technology to adapt the world to our purposes, and our own ability to adapt to technology is made possible by the way our brains perceive space. At the same time, our ability to step outside of the world's geometry mentally has resulted in a sad kind of detachment between us and the rest of the planet. This state of detachment may contain some important clues about another of the great paradoxes of human nature: how can a being whose mind is capable of such dramatic acts of understanding and creation wreak such havoc on its own home that its very future stands in some doubt? Perhaps the more pressing question is whether understanding where mental space comes from and what is done with it can help us to find solutions to such vexing problems. Can we rethink our relationship with space to make us more aware of the effects of our actions on the state of the planet? Can the artful design of buildings and cities encourage us to make connections to our spaces and places that will help us to take more responsibility for them? Could the secret to recapturing our sense of environmental stewardship reside in recapturing some of the connections with the planet that were possessed by ancient human forebears whose lives depended at every moment on their understanding of *where* they were? Can we co-opt some of the same technology that has helped us to conquer space in our personal lives to make us more aware of the wider sweep of space that is beyond the purview of our senses?

I hope you will find this book an optimistic one. It would be sad to think that a species that can produce Einstein, Mozart, Mother Teresa, and Shakespeare could short-circuit its own future, and the future of its home planet, because of a curious cerebral glitch that has simultaneously allowed us to conquer the wide spaces of the solar system but to become lost in a shopping mall. It seems more important than ever that we recognize that the most important impediments to our survival and flourishing are not technological barriers but psychological ones. More than anything else, moving forward in time and space will require that we understand not just who we are but *where* we are.

PART I

WHY ANTS DON'T GET LOST AT THE MALL

How Humans and Animals Navigate Space

CHAPTER 1
LOOKING FOR TARGETS

SIMPLE TACTICS FOR FINDING OUR WAY
THAT WE SHARE WITH ALL OTHER ANIMALS

Following the light of the sun, we left the Old World.
CHRISTOPHER COLUMBUS

We've all done it. At a meeting, a conference, a wedding, or a simple potluck gathering with friends, the food appears. Though manners may prompt us to restrain ourselves for a few minutes, our antennae wave, our restless feet shuffle, and we make a beeline for the tables. If a scientist were to hover above us and measure our movements, it would be easy to show the average guest-to-plate distance as a steadily decreasing mathematical function. This class of behavior, called taxis, is the simplest kind of spatial behavior that can be imagined. All that is required is a target (that magnificent roast of beef), a sensor or two (our well-tuned nostrils and eyes), and some kind of motive force (sore feet squeezed into formal shoes will do nicely).

Life does not always treat us so kindly, though. On our way to the table, Longtalker Larry makes a perfect intercept course. How to rearrange the missile trajectory so as to home in on the canapés while avoiding verbal entanglement with Larry? The buffet table has two rows of food. On the closest side is Aunt Betty's famous potato salad, but it looks a little bland. The better bet is Sarah's Spicy Potatoes, but they're just out of reach. We'll need to thread our way through a crowd, momentarily losing sight of the target completely, in order to plan the return foray to starch Valhalla on the distal side of the room. What's the quickest way? Perhaps the party is in a building we've never seen before. The sweet aromas are everywhere, but compared to what vision gives us, they don't make much of a spatial cue. Which way do we go first? How do we conduct an efficient search?

Compared with many of the stories of feats of navigation that I will relate to you, finding your way to and then around a table full of food is small potatoes (Sarah's if you're lucky). Nevertheless, all such behaviors, ranging from the trivially simple taxis to the complex wayfinding task, point to one basic truth of biology. Unlike the potted geranium sitting in my window, you and I, like all other animate beings, need to be able to move from one place to another to survive. In order to remain nourished, I must get up from my chair and go to the fridge to find food. In order to avoid a premature demise, I need to leap out of the way of the bus that hurtles down the road toward me. The whole raw biological point of my individual survival is to reproduce. But this, too, requires movement. In order to pass my genes on, I need to be able to get up and walk around until I find a mate. (This, you may argue, is something of an oversimplification.) To survive, we must come to terms with space and time. Whatever the physicists and philosophers might say about these things, movement is defined as a change in place over some duration of time. Given this, it is not at all surprising that nature has produced a wide array of mechanical devices that produce movement (legs, wings, fins, and so on). In addition, we have evolved an even more impressive arsenal of tools that allow us to know *where* to move—that is, to find our way through space to important goals such as sustenance, warmth, safety, and sex.

The simplest tricks of navigation are perhaps so obvious that we don't even think of them as being tricks. You are walking down the aisle in a grocery store when, just ahead of you, you see the box of spaghetti you've been seeking. With little or no conscious effort, the box is soon in your hand and then in your shopping cart. What's to explain? This seemingly trivial piece of behavior—moving to a clearly visible target—is something that we do hundreds of times a day. Such behaviors are required of all animals that move, yet they are accomplished in a wide variety of ways.

The most primitive kinds of animals, one-celled creatures such as bacteria, though their needs may be simple, must still possess a basic toolkit that allows them to find their way to conditions that sustain life: light, heat, and sustenance. Sometimes these unicellular denizens of our soil, water, and even our own bodies can employ a search strategy much like a child playing a game of blind man's bluff. Their rates of movement rise and fall with the activity of sensors tuned to the concentrations of heat, light, or chemicals that surround them, and these changes in movement bring them inexorably into contact with their goal. Other than the movement of a plant bending toward the light, it is difficult to imagine a simpler mechanism by which a living thing can deal with the problems of space.

In other cases, such tiny creatures as these may possess specialized equipment to help them guide their movements. In 1996, a group of scientists, headed by Dr. David McKay of NASA's Johnson

Space Center, claimed they had discovered fossil evidence for the existence of life on Mars in a lump of meteoric rock that had been collected from the Antarctic.[1] Analysis of the chemical composition of the rock left little doubt that it was of Martian origin, and the peculiar formations inside the rock looked suspiciously biological. Researchers thought they could see tiny cell bodies, reminiscent of our own earthly bacteria.

As some of McKay's early evidence has been disputed by others,[2] the initial excitement has died down, but he remains convinced that the particles of magnetite that were found in the sample once constituted a part of a Martian life form. Magnetite is found in various places on our planet, but one of the most interesting homes for this magnetic mineral is inside single-celled organisms that employ a unique style of navigation. So-called magnetotaxic animals use particles of magnetite as tiny compasses that orient their bodies with planetary geography. Though these magnetite bodies take advantage of the earth's magnetic field in exactly the same way that makes the Boy Scout compass face north, in this case it is not to help them to read maps correctly but to do something much simpler: the magnetite pulls these tiny aquatic animals downward into the lakebeds lining their watery homes, where they find food, safety, and comfortable temperatures. The origin of the magnetite found in McKay's samples is a matter that still swirls in controversy, but if he is correct, not only will his discovery constitute the first evidence of extraterrestrial life but his claim will be based on an elementary form of navigation.

The rudimentary navigational tools that I have described are based on a mechanism that allows an animal to drift up or down a gradient of light, heat, magnetism, or the concentration of some chemical. Such mechanisms can serve a variety of functions where animals need to get from where they are to an easily defined target such as a strong source of light or a warm pool of water. Simple as they are, some things are still not well understood about these elementary mechanisms. Indeed, some of the fine details of bacterial navigation have led researchers to suggest that these tiny beings possess a type of cognition not different in kind from that found in much larger multicellular animals.

When a hungry urban primate tries to zero in on Sarah's Spicy Potatoes in the buffet line, this is yet another form of taxis, but for reasons that will soon be clear, the technical hurdles that must be overcome to reach such targets are considerably more complicated than those faced by the average amoeba or slime mold.

THE POWER OF TWO

A frog sits motionless at the edge of a muddy stream, seemingly oblivious to the passage of time and the flow of events. When a fly happens within striking range, the frog's tongue lashes out to capture it with such speed and precision that the fly seems to have vanished into thin air. Clever scientific experiments using time-lapse photography have shown that the frog can not only discern the direction of the fly's movement but also assess the fly's distance with enough precision to ensure accurate contact between sticky tongue tip and hapless fly torso.[3]

Though prey catching in frogs might seem very different from taxic behavior in bacteria, what they share is that they are simple behaviors designed to help an animal make a connection with a spatial target. One advantage that an animal like a frog has over a microscopic one-celled critter is simply that of size. With a big enough body, sensors can be placed in such a way that they can be used to triangulate on the location of a target. A pair of sensors—the eyes in this case—can make precise estimates of the locations of target objects without having to engage in the complicated trial-and-error methods used by much smaller animals.

Bilateral symmetry (that is, a body composed of two more or less identical halves) is common in nature, and with such symmetry comes paired sense organs. The mechanism by which pairs of sensors can produce useful orienting behaviours can be exceedingly simple. A basement hobbyist can easily construct a small machine capable of such seeking behaviors using nothing more than a pair of sensors (for example, simple light detectors that can be purchased for a few pennies at an electronics shop), a pair of wheels, and a powered motor. By wiring the machine together in such a way that each sensor is attached to a wheel on the opposite side of the body, the machine can be made to roll rapidly toward sources of light. Alternatively, reversing the wiring will produce a timid machine that seeks out dark corners.[4]

More sophisticated uses of paired sensors involve comparing the images that are presented to each sensor to arrive at an estimate of the location and distance of a target. When we look at an object, its image falls in slightly different locations in each of our two eyes, and our brain can compute the distance of the object based on such differences. When we listen to a sound, the differences in the qualities of sounds arriving at our two ears can be used in similar fashion to compute the location of the sound source. The power of two in this case means that animals possessing paired sensors do not need to engage in hit-and-miss games of blind man's bluff in order to get close to the things they need. Instead, comparing the messages conveyed by each of the two sensors provides a rapid and accurate estimate of the location of a target. In simple machines built with light detectors and wheels, or in frogs and toads sitting stoically waiting for dinner to come within tongue's reach, the use of paired sensors is a considerable advance over the simple taxic mechanisms of bacteria. In more sophisticated animals like us, many more layers of neural machinery are involved in regulating our movements with respect to targets of interest. As preponderantly visual beasts, the story begins with our eyes.

———————

Spend a minute or two observing how your own eye movements contribute to your perception of the world. Find a point somewhere in the room and try hard to maintain your gaze on that location. While doing this, notice how much you can see of objects just outside your fixation point. If you hold your gaze steady, you'll notice that your perception of the rest of your setting consists of nothing more than a few blobs of varying brightness. Notice how little can been seen clearly when the eyes are held in a stationary position. Visual details are available in a small region of space around your fixation point, but nowhere else. To build an integrated view of the layout of the space we occupy, we need to move our eyes ceaselessly.

Working in the 1960s, when the technology for studying eye movements was primitive compared to the tools that are available to us today, Alfred Yarbus, a pioneer in the scientific study of eye movements, had participants in his experiments wear small mirrors that were attached to their eyeballs by means of small suction cups. (Yes, it was unpleasant. And, yes, Yarbus participated in his own experiments.)[5] In some experiments, participants were asked to examine paintings while Yarbus recorded the patterns of their eye movements. When the eye-movement recordings were superimposed on the paintings so that it was possible to see what the participants had been looking at, Yarbus discovered that eye movements were not scattered randomly across the paintings; nor did they seem to carry out any kind of systematic search (such as from top to bottom or from left to right, as one might imagine a machine would do). Instead, the eyes tended to seek out the parts of the picture that were most salient. For example, an inordinate amount of attention was paid to the eyes of the human figures in a painting. Yarbus was able to show that the pattern of eye movements seen during the viewing of a painting depended on the context of the viewing. If he asked people to answer questions about what they were seeing, their eye movements would reflect the strategies that they were using to search for answers. Our eye movements are not driven by what is biggest, brightest, or flashiest in a visual scene. They reflect the purpose of our looking.

Though Yarbus's clever experiments stimulated legions of future researchers to use measurements of eye movements as a kind of window into our minds, he was limited by the crude technology of his day. Participants were required to have their heads restrained for periods as long as three minutes while viewing his pictures, and the little stalks that were attached to their eyeballs were uncomfortable and distracting. Today, it is possible to measure eye movements with great accuracy using a much less invasive method. Participants in such experiments can simply wear a pair of glasses that contains miniature cameras to record the movements of their eyes. Using this method, much has been learned about how our eyes capture critical information.

While we move about, we use a series of quick glimpses, called fixations, interleaved with rapid eye movements called saccades. The average duration of a fixation is about half a second. Though there are slight variations, all saccades take roughly the same length of time, less than one-tenth of a second, regardless of the distance the eye travels during the movement. The greater the distance, the faster the eye moves. (Indeed, saccades are the fastest movement produced by the human body.) This detail is important because it suggests that saccades are programmed before they begin. In other words, before the eye begins to move, it knows where it is going. Generally, movements that have this property, whether they are movements of the eyes or of missiles loaded with nuclear payloads, are called ballistic movements.

These patterns of saccades and fixations have a definable structure to them, related to the actions that they accompany. Fixations vary in length depending on what they are for (locating an object, assisting in a movement such as grasping, checking something). These extraordinary patterns of fixation and movement are one illustration of the elegant *pas de deux* between perceiver and perceived. Our senses don't merely take in the world. In a way, we actually *make* the world we live in through these kinds of interactions. In the most superficial way, our movements through space may resemble those of bacteria and slime molds, but our progress toward the buffet table is underpinned by an elegant and beautiful perceptual dance that is largely beyond the reach of consciousness. With great concentration, as in the exercise I encouraged you to try earlier, we can become aware of the occasional eye movement or head turn, but we couldn't possibly have a genuine firsthand experience of the staccato visual sampling that underlies our stable perceptions of the visual world.

GRASPING SPACE

Movements such as reaching, grasping, and walking have been the subject of intense scientific scrutiny. One reason for this is that the study of such movements has much to tell us about how perception and movement work together, but another, more significant reason is the tremendous importance of our ability to grasp and manipulate objects. Everyone has heard the old saw that the main reason human beings have come to dominate the planet is our possession of an opposable thumb. Though this is a dubious claim (I would put my money on our massive cerebral cortex rather than on our thumbs), there is no doubt that our ability to coordinate our eyes and our hands to interact with the world with exquisite precision is a major hallmark of what it means to be a human being. A few other animals have impressive abilities to manipulate objects (raccoons, for example), but no other animal comes close to our combination of speed, precision, and flexibility in organizing skilled movements using visual control.

Though we reach for objects hundreds of times a day without a second thought, the problems that must be solved to complete these movements accurately are formidable. We must transform a viewed target location into a set of muscle contractions. If this seems easy, remember that the exact muscle contractions that are required will depend not just on the location of the image of the target on the retina but also on the position of the eye in the head, the head on the body, the arm on the shoulder, and perhaps even the orientation of the torso (think of bending over and picking up an object from the ground). In order to calculate the appropriate muscle contractions, it is crucial that our brains keep careful track of the relative positions of different parts of our own bodies as well as the appearance of the visual scene in front of us. We can do some of this work by using a specialized set of sensory receptors embedded in our joints and muscles. The outputs of these so-called proprioceptors report to our brain on the position of our body. In addition, whenever our brain sends a command to our muscles to move, a copy of that command is kept at hand in a neural filing cabinet so that we can use it to keep track of the expected consequences of each movement that we make. Our brain tries to save time by predicting the consequences of a movement before it has even taken place.

When we move our eyes, our hands, and our arms, we need only keep careful track of the relative movements of our body parts—eye relative to head, head relative to body, hand relative to shoulder, and so on. Walking changes everything. With each step, we take flight from the surface of the planet, and when we alight we are in a new location. It is no longer sufficient to measure our own muscular contractions or motor commands to determine our exact position in space. We need an entirely new set of tools.

Carrying a full glass of beer across a crowded barroom can be tricky business. When standing still, or walking at a smooth, unchanging speed, the beer sits securely in the glass, no tidal waves of liquid threatening the floor or our clothes. But each change of direction or speed can cause the precious liquid to slosh around in the glass. Now imagine an observant and scientifically minded drinker wandering across the floor with glass in hand. She might notice that the way the beer moves in the glass is related in a very orderly way to the movements of the glass. Sudden changes of motion cause predictable reactions in the shape of the surface of the ale. In fact, a careful observer could calculate the path of the glass through space by doing nothing other than measuring and recording such changes (though she might not be the most fun person to drink beer with). To calculate accurately, she would need to note each and every movement of the surface of the fluid. If she was distracted for even a moment, or if her memory failed her, the missing data would cause her to lose track of her position completely.

Many animals, including human beings, have a specialized set of organs that sense movement in exactly the same manner as our observant beer drinker. These structures, called the vestibular system, consist of a series of interconnected chambers and tubes within the middle ear. These wondrously shaped vestibules, looking a bit like a curvy architectural creation by Frank Gehry, are filled with a viscous fluid. Inside each of these tubes is a small chunk of gelatin, studded with tiny crystals of limestone to give it added weight. As our head accelerates and decelerates through space, the blobs of gelatin wobble around just like the beer in the glass. Tiny hairs embedded in the blobs are bent by each wobble, and these bending movements send signals to our brain.

The vestibular system works remarkably well for controlling certain types of movement. For example, our ability to maintain fixation on a visual target as we walk around or even leap through the air is largely brought about by a precise dialogue between our vestibular system and our eye muscles. But as a device for keeping track of our movements in larger-scale space, the vestibular system has the same weakness as our beer carrier. Errors creep into the mix, and those errors accumulate over time. Without any help from other sources, the vestibular system will become lost and disoriented. One possible source of help comes from the visual system, which has specialized abilities to keep track of our position as we move through space.

———————

Our understanding of how vision contributes to our perception of space and motion advanced when a newly minted researcher, James Gibson, co-opted to the U.S. Air Force during World War II, stood on a runway watching fighter planes landing.[6] Without question, landing is the most difficult part of flying—seasoned pilots will tell you that the definition of a successful landing is one that you can walk away from. In wartime, when new pilots needed to be trained quickly and in large numbers, there was a tremendous incentive to understand what made landing an aircraft so difficult. There was also great interest in developing a psychological test that might predict a person's aptitude for flying. Both of these problems fell to Gibson. Gibson must have been acutely aware of the personal consequences of failure. One of his predecessors had presented potential pilots with brief glimpses of the silhouettes of different types of aircraft and then asked them to identify the shadows. This very difficult task was an abysmal failure in predicting flying aptitude, and its inventor was discharged to the front lines. When John Watson, who later became an important figure in twentieth-century psychology for his theories of learning, was assigned the flying aptitude task, he found a way to pass the job along to a colleague, perhaps saving himself from the embarrassment of failure.

James Gibson showed more perseverance than his predecessors and eventually came to realize that good pilots kept track of their direction of movement, their altitude, and their velocity by taking advantage of certain regular patterns of visual motion that were produced by a moving observer. Gibson called these patterns optic flow, and he considered them to be at least as important to our sense of our own position as the signals we received from our vestibular system. As we move forward, the images of different parts of the world sweep across our retinas, but the region of space whose image enlarges most slowly indicates our direction of motion and our target. What this means to a pilot is that as his aircraft arcs toward the ground, the part of the planet's surface that appears to be expanding most slowly, called the focus of expansion, is his point of interception with the ground. One part of being a good pilot is developing an understanding of how such optic flow information can be used.

The patterns of visual motion that Gibson described guide our movements all the time. While driving a car, for example, we can gauge our direction of motion using the focus of expansion. In a similar vein, as we approach a target, we can calculate when to slow down and stop in order to avoid a collision using simple calculations based on measurements of optic flow. Our ability to avoid being struck by oncoming projectiles, such as knowing when to duck to avoid being conked by a baseball, is also based on these kinds of calculations. There is even some evidence suggesting that human beings, and many other animals, possess specialized neural circuits for detecting and responding very rapidly to these visual motions.

There is little doubt that Gibson was correct in his surmise that we use optic flow to complete simple orientation movements similar to those that can be observed in animals looking for light, darkness, warmth, or food. All of these patterns of visual motion, both those caused by our own movements and those caused by movements of objects in the world, could theoretically be used to compute our position and so help us to know our place in the world. As we will see later, the calculations that are involved can become enormously complicated, and it isn't at all clear that we can carry them out very accurately, especially when our movements take us on the complicated paths of travel that characterize our everyday behavior.

———————

The simplest kinds of problems in navigation involve nothing more than finding a way to decrease the distance between oneself and a target that can be sensed directly. As we've seen, these kinds of problems can be solved using nothing more than some basic sensors, a means of movement, and some biological wiring that joins the two together. For a one-celled animal seeking sustenance in a lakebed, a sowbug on its way to the dark, moist underside of a rock, or even a basic robotic device, things can be just that simple. Though we humans share these basic elements with all other animals, our guidance mechanisms are embedded in a much larger and more complicated system. Our ceaselessly moving eyes perch atop a complicated tower of flesh, flicking from one viewpoint to another in an elegant dance that helps us to put together an overall view of the world. The basic rules that get us from the street corner to the bus stop, or from the kitchen table to the front door, may not differ substantially from those used by bacteria, insects, or other simple beings, yet the detailed differences in how we use our senses to construct a sensory world will assume increasing importance as this story progresses.

Many of the everyday challenges of space may involve nothing more than finding a way to move toward a target that is clearly visible, but this is hardly what we think of as wayfinding. More challenging and interesting tasks involve seeking out targets that cannot be seen directly. Here we enter a new realm where we find positions by using the relationships among things, rather than the very simple changes in the apparent size, shape, and strength of sensory signals that characterize our use of taxic mechanisms.

CHAPTER 2
LOOKING FOR LANDMARKS

HOW WE SEARCH FOR THE INVISIBLE
BY USING THE VISIBLE

The philosophy of the school was quite simple—
the bright boys specialized in Latin, the not so bright in science
and the rest managed with geography or the like.

AARON KLUG

One of the worst jobs I ever had was poring through old life insurance records to discover the names and birthdates of children of policyholders so that the company I worked for could create a computer program to send out birthday cards to those children. Though the job was staggeringly dull, there was one saving grace. The office I worked in was near the top of a skyscraper on the outskirts of downtown Toronto, and from its south-facing windows, I was able to watch the construction of the CN Tower, until recently the tallest free-standing structure in the world. The highest parts of this tower were built using a magnificent Sikorsky Skycrane helicopter, an undertaking of such significance that the schedule of appearances of the machine was published in some local newspapers and broadcast on the nightly news. I had a front-row seat, free of charge, provided I could master the art of pretending to fill computer coding sheets with names and dates while watching the tower take shape.

As a young man whiling away his hours at a boring job, I had no sense of the transformative effect the tower would eventually have on the city. The main justification for the structure was that the boom in high-rise construction in downtown Toronto had begun to impede various kinds of radio telecommunications. But the rationale clearly had as much to do with establishing a "world-class" landmark for the city, an identifiable icon of space-age advancement, as it had to do with the pragmatics of transmitting radio waves and microwaves. But as well as serving as a landmark in the more colloquial sense of the word—as a structure whose silhouette has become identifiable as a part of the Toronto skyline as readily as New York City's Empire State Building or Seattle's Space Needle—the CN Tower has come to serve as a true landmark in the navigational sense. Wherever you are in the core of the city, or even in the outer fringes, it doesn't take much of an effort to find the tower and thereby to help fix your own location. (The positional fix is helped along by the fact that the tower is located near the north shore of Lake Ontario, so one is very unlikely to be south of the tower.) The tower can also be used to gauge one's distance from the downtown core. When I drive into the city, along a highway that skirts the edge of Lake Ontario, the easiest way to judge my progress is to take a fix on the apparent size of the tower.

Used in this way, the CN Tower is a classic navigational landmark. Though the tower itself is most often not our final destination, we can find our goal by its relationship to the tower. We have gone slightly beyond the realm of navigating toward targets that we can see. We are no longer following plumes of the delicious aromas arising from buffet tables, or flying airplanes on to clearly visible runways using tools not conceptually different from those employed by everything from the *E. coli* bacteria in our guts to rattlesnakes hunting down field mice. Now we are using the visible to find our way to what is invisible. To do this means to have at least an implicit understanding of the spatial relationships between things. Such abilities place us in slightly more rarefied company.

Some of the first conclusive studies demonstrating animals' use of landmarks for navigation were conducted by the biologist Nikolaas Tinbergen. Though he was eventually awarded the Nobel Prize for his studies of animal behavior, Tinbergen was the black sheep in his family. In contrast to his industrious brother, Jan (who also won a Nobel Prize, for economics, in 1969), "Niko" was known for spending dreamy summer vacations observing and photographing animals rather than applying himself in a rigorous manner to any of the established branches of zoology.[1] What was most remarkable about Tinbergen was his flair for close observation of the behavior of animals followed by elegant and convincing field experiments that highlighted the principles underlying his observations.

On one family vacation, Tinbergen spent some time observing digger wasps. These wasps dig small nests in the ground, which they provision with captured and paralyzed insects so that when their young hatch they will have a larder awaiting them. This set of behaviors requires the wasps to repeatedly leave the nest and then return to it. Tinbergen wondered how the wasps found their way back to the tiny, almost invisible entrances of their nests.

Tinbergen's approach was characteristically simple yet effective. Wondering whether the wasps might be using visual landmarks to locate the nest entrance, he simply removed some of the natural objects that lay scattered near the entrance, sat back, and waited to see what would happen. When the wasps returned, it was clear to him that they had become disoriented. Tinbergen's experimental coup de grâce came from a further experiment in which he seized control of the situation by replacing the objects surrounding the nest entrance with a carefully arranged circle of pinecones. Once the wasps overcame their confusion and regained their ability to find the nest, Tinbergen shifted the ring of cones to a nearby location. When the wasps returned, they searched for the nest entrance in the center of the displaced array of cones, demonstrating that Tinbergen had correctly identified the manner in which wasps found their way home.[2]

This simple, informal style of experiment has come to form the backbone of studies of landmark navigation in animals ranging from the wood ant to the human being. Today, though the sophistication of the methods used has advanced to a state that Tinbergen could not have imagined, the logic of the experiments has changed very little.

Tom Collett, an experimental biologist at the University of Sussex in England, has spent a lifetime studying spatial navigation in insects.[3] Collett's taste in species has been democratic—he has worked with a variety of ants, bees, and wasps, so his findings generalize well. One question that preoccupies Collett is the way insects memorize configurations of landmarks. What do they look for as they try to return to the nest? Despite the small size of their brains, insects show surprisingly sophisticated cognition. When leaving a place to which they need to return, flying insects carry out highly structured orientation flights, in which they turn to face the goal and carry out a series of swooping arcs around it. Their purpose is to produce a kind of image or snapshot of the vista surrounding the goal location that can be recalled later.

Insects that must travel over long distances before returning home may carry out several of these orientation exercises in order to memorize the sequence of landmarks they should observe on the path to home. Insects use these snapshots by moving about until they are able to match a view of the terrain that they face with a memorized view, almost as if trying to find the right puzzle piece to complete a picture. Collett discovered this behavior by using methods much like those pioneered by Tinbergen. By manipulating the landmarks that surround the target, one can force insects to make predictable errors. For example, by replacing the landmarks seen on orientation flights with other landmarks that look similar but are larger, one can fool insects into thinking that they have arrived home early. This is because the insects misconstrue the larger landmarks as being closer to them than they really are. This is exactly the same principle we use when we judge the size of a building to orient ourselves. The closer we are to it, the larger we expect it to be.

Tinbergen rushed through his doctoral thesis based on the digger wasp experiments because he and his young wife were anxious to set off on a new adventure. Tinbergen had found an opportunity to accompany a small Dutch expedition to Greenland for the International Polar Year in 1933, where he ended up staying for fourteen months living among a small group of isolated Inuit who maintained a largely traditional lifestyle. While Tinbergen's ostensible role as a biologist was to study several Arctic species, including husky dogs and snow buntings, there is no question that the Inuit community where he lived was a natural match for his careful, observational approach to nature. These people, having survived for thousands of years in forbidding environments, lived by their wits, foresight, and, most important, their keen sense of observation.

There are different regional specializations among Arctic navigators, but the Greenland Inuit, like their Baffin Island cousins, live in regions where there are normally plenty of environmental landmarks, such as the cliffs that form the walls of the steep valleys and fjords. Such landmarks can be seen from great distances and, even when bad weather makes visibility poor, it is usually possible to cling to the valley wall and to keep it in view so as to avoid becoming lost.

A few years ago, on a visit to the community of Clyde River on Baffin Island, I experienced this form of navigation firsthand. On a day trip to explore the stunning beauty and rugged terrain of a nearby fjord, my group traveled over sea ice by snowmobile for about 50 kilometers into a deep valley. When the skies suddenly clouded over and the air filled with the foreboding heaviness of a blizzard, we began a nervous retreat. As the gloom closed in, we lost so much visibility that it was no longer possible to see a horizon or to make any distinction between land and sky. In spite of this, we never lost sight of the difference in color between the rock face on our left and the open ice on our right. Although unnerved that I could see nothing other than the thin contour between brown rock and white ground in my left periphery, I knew we were in no danger of becoming lost provided we maintained visual contact with that dark ridge. I was so used to living in overdetermined visual environments teeming with navigational cues that I clung to my visual lifeline like a child clinging to his mother's hand.

My sense of danger was considerably heightened by the possibility that, as far out as we were on the sea ice, we might encounter a polar bear. The head of the local Royal Canadian Mounted Police detachment had not helped when, before we set out, he had carefully explained a deep truth for dealing with bears on the ice. With a somewhat concealed twinkle of amusement, he had told me to keep my eyes open and to always remember one thing. "If you're out on the ice away from your snowmobile and you spot a polar bear, remember the thirty-second rule."

I bit. "Thirty-second rule? What's that?"

"In thirty seconds, you'll be dead."

When we caught our first glimpse of the village, there was a palpable spirit of celebration. Having come close to a loss of spatial orientation that could have led to a quick death, I had never cherished my sense of place so dearly.

———————

One reason the Inuit are such excellent navigators is that they have an exquisitely tuned ability to pay attention to the visual features of objects and scenes. One oft-cited study, conducted in 1996 by psychologist John Berry, compared the visual abilities of Inuit, modern urban Scots, and the Tenne people, an African agricultural society.[4] On a small battery of standard psychological tests of visual function, the Inuit proved equal or superior to the Tenne and Scottish group on all measures. For the Inuit, survival on the land is so dependent on careful visual attention to detail that it is embedded in their language. Inuktitut, one of the languages of the Barren Inuit, contains what have been referred to as obligatory localizers. Just as some languages, such as French or German, require that the gender of nouns be specified and embedded into the syntax of a sentence, Inuktitut requires that the location and orientation of an object be specified as a part of the grammatical structure of a sentence. One example is the three-word Inuktitut sentence "Ililavruk manna ilunga," which translates as the twenty-word English sentence "Please put this slender thing over there crosswise on that end of that slender thing to which I am pointing."[5]

Not only do the Inuit cultivate such sharp observational skills that they possess linguistic specializations obliging them to notice and include spatial references in simple statements about objects but they also use a rich vocabulary to describe the land. Every small feature, hill, or outcropping of rock has a name. The names are woven into stories of events that took place there or describe objects they resemble. The cliff that I followed on my narrow escape from the closing blizzard was called Naujaaraaluit. *Nauja* is a seagull nest. *Raaluk* is a size modifier (big seagull nest). *Uk* changes to *uit* to denote a place. Naujaaraaluit is the place of the big seagull nests.

By naming landmarks and embedding them into stories, Inuit trekkers are using a tactic analogous to one that is used by digger wasps, but this tactic is much less dependent on geography. The wasps memorize routes using orientation flights in which they simulate their paths to a goal by carrying out the same set of flight movements that they will use later to return to a target. While these movements are taking place, the wasps are recording in memory the appearances of key landmarks. Though skillful Inuit navigators may be just as successful at finding their way home, there has been a key shift in tactics. An Inuit explorer can sit inside by the fire recounting a story to himself and his clan that can have some of the same net effect as the wasp's orientation flight. This pattern—a human shift from the use of space and geometry to navigate long distances to one based on a mental landscape of words, stories, and ideas—is one that we will see repeated often in the pages to come.

No discussion of Inuit navigation can ignore the prominent built landmarks referred to as inukshuks. These stone sculptures, constructed to resemble human figures, are usually built on high ground with their arms pointing toward shelter. Occasionally, they are used to indicate productive fishing grounds, with their distance from the water's edge meant to approximate the distance at which fish are to be found. Sometimes strings of inukshuks are built such that peering through a hole in the middle of one can lead to a sighting of another. In addition to their role in navigation, inuk-shuks can be used as icons to represent departed family members. They can also serve as a kind of hunting aid, like a scarecrow, influencing the direction of movement of caribou herds. Though these structures are certainly used by Inuit for navigation, they have functions that go well beyond those of simple landmarks. Inuk-shuks are embedded in the cultural fabric of the Inuit.

This is an interesting extension of the human penchant for carrying wayfinding directions in our heads in the form of words and stories. In the case of inukshuks, physical symbols of the stories and legends are placed directly into the picture using carefully balanced piles of rocks. We mark up our environment in ways that adapt it to the mental toolkit we use to find our way through it. In Inuit culture, the connection between land, story, and inuk-shuk symbol is fairly straightforward. In modern urban cultures, our uses of landmarks to glue culture and memory to place may be much more dramatic, but the impulses that cause us to create such connections may be universal, with origins in the unique nature of our cognitive makeup.

In recent times, a powerful demonstration of the power of a landmark concerns something that is really the inverse of my experience of watching the growth of the tower in Toronto. On September 11, 2001, the world watched as the World Trade Center in New York City was shattered by the murderous act of a terrorist group. The Twin Towers, initially accepted with great reluctance by New Yorkers, had eventually come to be seen as emblematic of the city's image, in much the way that their architect, Minoru Yamasaki, had envisioned. Not only could the towers act as an explicit navigational beacon for confused travelers but their iconic value in identifying the skyline of the city was beyond limit. When the towers were destroyed, the response was experienced globally as a combination of disbelief, grief, and, among the local population especially, a yearning for what had been lost so strong that it produced an obsessive tendency to visualize or even reproduce the iconic form of the towers. The most dramatic example of this tendency appears in the annual "Tribute in Light," in which powerful spotlights shine into the sky from south Manhattan as a reminder of what was lost. To this day, it is almost impossible for the eyes not to be drawn to the former location of the towers as one flies into the city or observes the skyline from across the waters in New Jersey.[6] It is a tremendous tribute to the power of the human mind to organize space by attaching it to our stories that the hole left in the landscape by what has been lost can also help to connect us to a place.

HOW WE USE LANDMARKS TO SEARCH FOR WHAT WE'VE LOST

My wife, though she is able to keep track of the locations of all of our children, our social schedule, the birthdate of everyone she has ever met, and all of her secret hiding places for cookies that she doesn't want me to eat, is utterly incapable of keeping track of her car keys. One of her more notorious episodes of key loss took place in a municipal park with a large grassy playing field. In the middle of a game with the children, she put her keys down in the grass and walked away. At the conclusion of the game, a mad hunt for the small fob of keys in the large field ensued. The experience will be familiar to many. We walked around, trying to recall which particular features of the landscape we had noticed in the area where we had played. Our intuition told us, correctly, that the smaller the distance between such landmarks and the possible location of the keys, the more likely would be our success. If we could only find a way to identify the clumps of grass that we'd been near during our game, it would be much easier for us to find what we had lost. But alas, there were no such easy landmarks, and the keys were never found.

Looking for a lost ring on a beach, gloves forgotten in a shopping center, or a wayward screw in a basement renovation project, we are most likely to take stock of our surroundings and to try to narrow down our search location until all visible landmarks match the orientation and distance that we remember seeing when the lost object was last in our hands. When such strategies fail, it is most often because our memory of the appearances and locations of the landmarks has let us down. We are more likely to succeed when we have carried out a similar search in a similar location on many previous occasions, just as the digger wasp has made many round trips between nest and field to provision the larder for its offspring.

A clever experimenter can ask detailed questions about how we use landmarks in such cases. For example, imagine that, like the digger wasp, you have learned to look for one particular location that is defined by its relationship with a series of closely spaced landmarks. Over the course of thousands of trials, athletes learn to place a hand or foot, throw a ball, or place a stick into a particular location based on its relationship with a set of visible landmarks. What would happen if the relationships among the landmarks themselves were changed? Imagine that your target is always placed in the middle of a square whose corners are marked by four orange pylons. Without your knowledge, a devious experimenter comes along and moves all four of the orange pylons to make the square larger. Suppose that the change is small enough that you fail to notice it. Where do you think you would look for the target?

The answer seems to depend on what kind of animal you are. In experiments that are exactly analogous to the situation I've described, some insects and some other animals (rats and gerbils, for instance) will search in a series of four different locations, each related to one landmark, and each one the same distance and direction from that landmark as before the change. It is as if the animal has memorized the exact geometric relationship between each of the landmarks and the target, so it searches four places, each one the predicted target location according to the position of one of the landmarks.

Human beings, on the other hand, continue to search in the center of the square as defined by all four landmarks, even though the distances between them have changed. It is as if the "code" for the location of the target is made up of the relationship between the target and *all* of the landmarks, rather than any one of them.[7]

It isn't completely clear what these differences in behavior might mean, especially since other animals, such as certain birds, also treat landmark arrays in the human fashion. But one possibility is somewhat related to the Inuit tendency to navigate according to named landmarks connected by stories. Suppose that the four orange pylons in our hypothetical experiment were mentally encoded not as a set of four discrete objects but instead as a single shape—a square. We could even imagine that this shape is encoded in our memory not as a very specific collection of corners of a particular size and in a particular location but as something much simpler—like the mental equivalent of the word *square*.

This kind of encoding would have a couple of advantages. For one thing, it would ease our memory load considerably. But in addition, it would allow us to make predictions about how the shape of the collection of landmarks might look from different points of view without our necessarily *being* at those viewpoints. Like an Inuit navigator describing the route through a storm-ridden Arctic fjord from the comfort of the fireside, we could describe the location of the target as being at the center of the square. The whole geometric problem is boiled down to, in this case, two simple ideas—center and square. Encoding locations in this highly schematized form, though it has some advantages for memory load and mental portability, also has liabilities. Because such forms jettison geometry, we are prone to make errors when landmarks move around. What if it *is* the case that one or two of the landmarks we've been using have moved? How do we then find our target? In such cases, a rat might have the advantage over us because one of the landmarks could still be used to define the position of the target. For us, on the other hand, if the four landmarks no longer define a nice square, we are left to guess where the center of the square used to be, and any guess would lack precision.

DISTANT LANDMARKS, DIFFERENT RULES

Recently, I visited a colleague in Santa Barbara at one of the campuses of the University of California. I had never been to this campus before, and so my friend sent me directions to his office. The first part of his instructions consisted of a short list of commands to be followed while driving. Provided that I paid attention to the road signs listing highway interchanges and followed the right sequence of left and right turns, I was assured of landing in the correct parking lot at the university. The crucial part of his directions dealt with what happened to me after I left the car. He instructed me to leave the parking garage and then walk so that the mountains were behind me and the ocean was in front of me. In a way, this seems to be entirely in accord with everything we discussed in the last chapter—I was following a simple taxic strategy by using the ocean as an attractive target and the mountains as a repulsive one. In spite of appearances, though, there is a crucial difference. The real target of my movements was not the ocean but my friend's office. In other words, I was using the mountains and the ocean in an entirely different way—these large, obvious, and clearly visible objects were serving as landmarks for a target that, from my starting position, was entirely invisible. Fortunately for me, the man I was meeting was an expert in spatial navigation, so his directions, efficient and precise, guided me to his office with laser accuracy, almost as if I were performing like a small insect on a homing mission.

When I used landmarks to find my colleague's office in Santa Barbara, I didn't need to think about the shapes of collections of landmarks at all. I got out of the car, set the mountains behind me and the ocean before me, and I walked. Why are situations such as these so different from those such as looking for lost keys in the grass? In part, because the landmarks involved are very large and very distant. These types of landmarks, sometimes called distal or global landmarks by scientists interested in navigation, behave differently from the clumps of grass or orange pylons that we have been discussing.

Distal landmarks would not work at all to help locate a lost wedding ring on a forest floor, but they *can* help a monarch butterfly find its way from a summer home in Canada or the United States to overwintering sites in Mexico, or a savvy hiker find her way out of a dense forest and to the nearest highway. Landmarks like the sun, moon, distant mountain ranges, or even skyscrapers or cityscapes enjoy special status among the tools used by navigators both human and otherwise because they have an ideal collection of properties that can assist wayfinders. If they are both very distant and visible, then they are also probably very large and therefore immobile. Large, immobile objects can be relied upon to stay where they are, and so define locations and directions. (The sun, moon, and stars are exceptions. Such objects change position as the earth rotates, but because such movements are predictable, we can learn their patterns and use these objects as landmarks anyway.) The other kind of landmark—pinecones near a wasp nest, clumps of grass, or bits of rock in a meadow—are not nearly as reliable. They might remain in place, or they might be kicked away, eaten, or even disturbed by curious scientists.

Many animals have such an implicit understanding of the differences between local and global landmarks that these can be shown to play different roles in their lives. Small, local landmarks are quickly disregarded when they show signs of being unreliable. Cues derived from large, distant landmarks are clung to with much more tenacity, and some studies show that animals are more reluctant to abandon representations of space based on the orientation of such masses. Determination may move mountains, but it doesn't happen very often.

———————

Some of the most exalted of human navigators have learned to conquer vast tracts of what, to the untrained eye, can appear as the archetypal example of flat, featureless territory—the open sea. Long before there were global positioning satellites, maps, or even compasses, human beings found ways to complete long voyages over open water employing the most subtle of cues, including landmarks.

Puluwatese marine sailors of the South Pacific routinely embarked on voyages of up to 650 kilometers in large, wind-driven outrigger canoes, navigating completely without instruments. In many cases, these expeditions were conducted for the simplest of reasons— to obtain a new stash of tobacco, or even for the men to escape the pressures of village life for a few days by fishing on a remote island.

Among the Puluwatese, the navigators were a revered class. They underwent rigorous secret training and a long apprenticeship under a master navigator before being allowed to guide a crew on the high seas. Much of this training involved painstaking memorization of the sequence of appearances and disappearances of stars along the horizon as an evening's sail progressed. Although the path that stars take across the heavens changes from season to season, where they appear and disappear is more stable.

Most voyages consisted of a series of segments connecting islands that were close enough to be visible from one another, but, given the distances involved, target and landmark islands could be tiny brown dots in a vast blue ocean. Puluwatese navigators learned clever strategies that increased the visibility of these distant objects. Seabirds often aggregate near particular islands, and so their appearance in the sky foretells the sighting of a landmass. By attuning themselves to the identity and habits of such birds, Puluwatese navigators learned to see over the horizon, into the future. Landmasses, even small ones, affect patterns of cloud formation. As well, land and water have different effects on the appearance of the sky that overarches them. Skilled navigators learned to tune keen vision to these subtle signs so that they not only could detect an island before its silhouette crossed the threshold of visibility but could often identify it based on the reflections cast by its trees, bushes, and lagoons on the sky above.

Puluwatese navigators also devised some subtle conceptual tools that could help fill in the blanks left by the organization of the human mind. They tracked their progress at sea using a method called *etak,* which involves lining up a succession of stars near the horizon with the positions of visible landmarks. The navigators memorized these star positions during their training, and their later observations allowed them to estimate their position in the sea. Interestingly, though Puluwatese navigators learned to master complicated lists of star–landmark combinations that could help them to gauge their progress, they apparently had no real understanding of the geometry of space that made these feats possible. Try as he might, Thomas Gladwin, an anthropologist who studied the Puluwatese people, could not make them see the geometric relationship between vessel, island, and star as it might appear to an overhead observer.[8] Gladwin reports, somewhat mysteriously, that the Puluwatese were able to line up star positions with the locations of landmarks even when the landmarks were not visible. Although how is not clear to me from his account, my surmise is that they were somehow able to bootstrap the estimated location of an *etak* landmark from the sum of all the information and intuition that they were able to bring to bear on their current location, perhaps also using a little intuition and blind luck.

The navigational feats of the Puluwatese are almost beyond the imagination of any but the most experienced Western sailors. It is hard to comprehend the skill of a navigator who can be blown off a capsized canoe in an ocean gale in the middle of the night and not only regain his purchase on board the vessel but also manage to compute his location and bring his crew safely home. Such sensational feats of navigation, like those of the Inuit, were mostly based on a keen observational eye and arduous practice in the patterns of star movement and the locations of visual landmarks. Such knowledge was probably so deeply ingrained through training as to have become completely automatic, probably seeming almost mystical to the Western anthropologists who studied them. There is at least a hint of mystic reverie in the description given by David Lewis of the navigational prowess of one of his main informants, the indomitable Tevake:

> It was for eight solid hours that Tevake stood on the fore-deck . . . gazing intently at the sea and only moving to gesture from time to time to guide the helmsman. Then around 14.00 something more substantial than mist loomed up through the murk fine on the port bow perhaps two miles off. "Lomlom," said Tevake, with satisfaction. Very soon afterwards Fenualoa also became visible to starboard and it was apparent that Tevake had made a perfect landfall on the middle of the half-mile-wide Forest Passage between the

two, after covering an estimated 45 to 48 miles since his last glimpse of the sky.[9]

Reading this passage and wondering about the abilities of human beings so acutely tuned to their environment as to appear to use methods beyond the grasp of mere mortals reminds me of a conversation I once had, somewhere near the end of a long bottle of rum on an idle summer day, with a retired Nova Scotia fisherman. He told me that one of the most important skills for a fisherman was to be able to return to the same spot in the ocean from one day to the next. One didn't necessarily want to employ marker buoys, as these would also be visible to competing fisherman, so it was better if one was able to use fixes on distant shoreline landmarks to estimate position. I asked him whether he could also return to a good fishing position when visibility was poor and shoreline landmarks were invisible, and he assured me that he could do so. When I asked him what the secret was, he fixed me with a surprised stare as if he had suddenly discovered that he was dealing with a simpleton.

"You just know when you've gone too far." All too often, I wish this were true.

In the scorching heart of Australia, water is scarce, transient, and unreliable. Anyone hoping to survive in such an environment needs to know where to look for the precious stuff. Among the earthly environments where survival depends on mastery of physical space, the Australian Outback is one of the most formidable examples. We might expect that human cultures that have adapted successfully to such a bleak setting would, like Puluwatese or Inuit navigators, have mastered some special abilities to keep track of their locations. In one intriguing study, an Australian perceptual psychologist assigned Aboriginal schoolchildren living in the great Western Desert in Australia and young Australians of European heritage who had grown up in the same area a task that required them to memorize the locations of an array of natural or artificial objects.[10] Not only did Aboriginal children show superior recall but their approach to the problem was different from that observed in the European children. Aboriginals stared quietly at the array of objects before being asked to reconstruct their positions. The other group fidgeted, verbalized, and audibly ruminated on the task during the learning phase and often rushed to identify the locations of the several items that they had managed to memorize as soon as the task began. There is no reason to think that there might have been biological differences between the children of different backgrounds in this study. It is much more likely that the native Aboriginal children, having been raised in a culture that placed high value on knowing the places of things and their spatial relationships as a matter of survival, would find the solutions required of them in these psychological experiments to be completely natural.

In his remarkable book *The Songlines*, Bruce Chatwin proposes that there is a vital link between the stories that form the oral tradition of the Aboriginals and the size, shape, and appearance of their physical landscapes.[11] This link is thought by Aboriginals to hold the key to understanding Creation. Their creation stories suggest that the landscape was literally *sung* into existence during the first days, called the Dreamtime, and that the human occupants of the land, in their role as shepherds of all that has been created, are required to continue to participate in this song in order to help keep the earth's spirit alive. The shapes of the sounds in the songlines bear a tight relationship to the physical features of the landscape. Hilly land connects with undulating tunes. Flat areas connect with long, legato phrasing. Because of this, not only do the songlines play a role in the spiritual life of the traditional Aboriginal but they serve as navigational aids as well. Just as the Inuit embed a detailed verbal map of the physical landscape in their stories, the Aboriginal songlines, by connecting different parts of the landscape into a creation narrative, help people to find their way from one sacred site to another.

―――――――

Members of societies with ancient origins who depend heavily on being able to find their way over long distances have developed sensitivities that have allowed them to compensate for lacking some of the tools possessed by other animals. Insects, birds, mice, and rats may be able to learn directly the size, shape, and geometry of the places where they spend their time, but we humans are more likely to learn and memorize names, concepts, connections, and stories. One key to accurate navigation in human beings is the development of an acute sense of observation, so that even subtle signs of direction and distance can be used to help one find one's place. But in addition to skills of observation, early human societies developed rich oral traditions in which their knowledge of the land was embedded in stories involving creation, spirituality, and the life histories of their deities. At the very least, these oral traditions have helped such societies to come to terms with the enormous memory load required to maintain a detailed map of one's local environment based on landmarks.

Everyone has a warehouse of memories filled with landmarks that evoke powerful feelings and detailed stories from our pasts. When driving to Toronto, my first glimpse of the CN Tower never fails to remind me of the wasted days spent working at that insurance company as a student, but it also somehow becomes wrapped up with many other memories of adventures, escapades, and close calls in the city I called home as a child and teenager. Similarly, on my first trip to New York City, the first glimpses of major landmarks such as the Statue of Liberty and the Empire State Building were what brought me to full frontal contact with that storied city. When I saw the statue, I knew where I was. When we travel to new places, we gravitate first to those epic places, the grand sights that appear on the first postcards sent home—the Eiffel Tower, the Great Wall, and the Bridge of Sighs. We seek these places out less because they serve as convenient navigational beacons (though, significantly, they often do) and more because standing near such monuments helps to stamp our presence in these new locations into memory. We may laugh at the excited tourists having their pictures taken in front of the Disney Castle in Orlando, but everybody understands the impulse. Where no landmarks exist, we may even make our own by etching initials, symbols, or epithets into rocks or trees, returning to visit these sites again and again throughout life.

When I visited Walden Pond for the first time, I picked up a pebble from the shore and put it in my pocket. Months later, when I found myself leaving a village I had come to love as my home, I left the pebble behind, as if to weave the story of my life and attachments into the terrain, connecting space and place. We modern folk may lack songlines or elaborate creation stories, but we do what we can to weld ourselves to the landscape with whatever tools we can devise. A part of this impulse must surely arise because, deep in our bones, we understand the need to belong to particular places. Landmarks, because they locate us, are integral to the fulfillment of that need.

CHAPTER 3
LOOKING FOR ROUTES

HOW WE TRY TO KEEP TRACK OF WHERE WE ARE
BY NOTING WHERE WE HAVE BEEN

The wonder is always new that any sane man can be a sailor.
RALPH WALDO EMERSON

In an ancient Greek legend, King Minos of Knossos commanded his brilliant engineer, Daedalus, to build a safe confinement for the Minotaur, a hideous beast born of an illicit union between Minos's wife and a white bull thrown from the sea by Poseidon. Daedalus built a labyrinth—an enormous cavern filled with passages of vast complexity and almost impossible to navigate. Not only did this elaborate construction keep the citizens of Knossos safe from the Minotaur but it provided Minos with a good outlet to vent his spleen over the death of his son, Androgeos, who had been murdered while traveling in Athens. To avenge Androgeos's death, Minos had Athenian virgins shipped to Knossos and sent into the labyrinth for the pleasure, and sustenance, of the Minotaur.

This ritual continued until Theseus, the brave son of the Athenian king Aegus, persuaded his father to allow him to cross the sea to slay the Minotaur. Theseus traveled to Knossos, dispatched the beast, and found his way safely out of the labyrinth. Theseus had some help. Ariadne, Minos's daughter, provided a long length of thread that Theseus unwound behind him as he made his way through the labyrinth, allowing him to find his way back to freedom.

The story of the labyrinth, the Minotaur, and Ariadne's thread finds resonance with us because we understand implicitly that we have a fragile grasp on place, and because of our deep human fear of becoming lost. The thread that ultimately saved Theseus's life was a substitute for a missing piece of navigational equipment: we often become lost because we cannot keep track of our position by remembering our own movements. Some of the most dramatic stories of our failures of navigation come from accounts of the experiences of wilderness explorers, and some of these come about when we have no equivalent of the priceless thread provided to Theseus by Ariadne. The very worst cases usually come about when we are either unable or unwilling to make allowances for our own fragile grip on space.

Edward Atkinson, surgeon and base camp commander on Robert Scott's ill-fated Antarctic expedition of 1910–1913, pushed at the door of his warm cabin and peeked through the crack into a blizzard. His destination, a scant 30 meters from the cabin, was a set of weather instruments. Though the readings seemed a foregone conclusion, the data had to be logged, and it was his turn. He looked at the stacks of cold-weather gear groaning from hooks near the door. Shrugging his shoulders, he pulled wind gear over his clothes, leaving his heavy coat and warm fur-lined boots where they lay. He wanted to finish quickly and return to his bed.

He pushed his way outside. The wind slammed the door shut behind him. In two steps, he was enveloped in swirls of blowing snow. The instruments lay straight ahead, but he could see nothing through the snow. He staggered forward, counting steps. In fair weather, the 30 meters would pass with the same number of paces. Walking into headwind, eyes closed against the stinging pricks of driven snow, it was more difficult to judge. Two hundred? Three? Amid the shrieking confusion of the storm, Atkinson lost count. Deciding he had gone too far, he turned around. The wind, constantly changing direction, knocked him to his knees. He crawled forward, feeling his way ahead, hoping to grasp rocky landmarks. Time slowed. His body first chilled and then began to heat up. He felt the burn of hypothermia and resisted the urge to throw off his hood and open buttons. His hands found resistance against a rocky ledge. He followed the ledge for as long as he was able. He felt unconsciousness closing in and staved it off by repeating short phrases to himself, trying to conjure images of family and friends. A fatal drowsiness crept into the worn edges of his consciousness. He kicked an opening in the side of a snowdrift. Without his warm boots, he would lose his feet, but he thought if he could get out of the wind, he might live for a few more hours. Lying back, he felt resignation creeping into his numb body, ready to accept death if this was his time. As if it were responding to his acquiescence, the wind let up slightly. Through slits of half-closed eyes, he saw a hazy glow above him. Mistaking it for a welcoming angel, he raised his head. Beneath the lonely confusion of deep hypothermia, he was able to make out a few bold contours. The storm was ending. The light was from the moon. Atkinson summoned the last traces of his will and dragged himself up the low hill to the cabin. He had been out in the storm for four hours.[1] (He did not lose his feet.)

Considering the adverse circumstances—the changing wind direction, the blinding snow, and his inadequate protection from the cold—Atkinson's quick and profound descent into disorientation is not surprising. What is more interesting is his lack of circumspection in thinking that he could find his way down the barren slope to his instruments and back to his cabin without the proper use of his senses. Atkinson's attitude on this frigid, wind-blasted night was deeply symptomatic of one major conceptual shortcoming of Scott's entire expedition: these incredibly brave men, determined to reach the South Pole by whatever means, set out without having taken heed of the physical and mental hardships that were in store for them. Not only did they have inappropriate equipment (ponies instead of dogs) and training (they soon discarded the skis they had brought with them, not knowing how to use them) but they lacked an appreciation for the fragile grasp that we human beings have on our sense of *implacement,* our knowledge of where we are, our spatial relationship with the objects that surround us, and the movements we need to make to reach them.

You and I are scarcely different. Think back to the last time the lights failed in your home. Remember your creeping movements from wherever the darkness found you, along walls and around furniture to where you thought you had last seen the flashlight or the candles. Remember the doorway encountered sooner than expected, the head bumped on the strangely misplaced lamp, the otherworldly sensation that you were stumbling around in a space that resembled your home but in which all scale, proportion, and position was askew and unexpected. Even when we know where we began, we do not cope well with problems that require us to correlate our movements with our position in space.

There is a paradox here: how could it be that a member of a species that has found ways to map and travel far into space can find himself completely lost after a few seconds of wandering through a snowstorm or even a dark bedroom? How could an animal capable of solving the problems required to get itself across many thousands of kilometers of land and ocean to within a few days' walk of a pole be so inept as to stumble, fall, and almost die a few meters from safety?

Given our clumsiness in the deeply familiar spaces of our darkened homes, it is not surprising that we human beings find many other ways to lose ourselves quickly, and sometimes with disastrous effects. Some telling statistics come from studies of "lost person behavior" that are designed to help guide search-and-rescue workers to those who wander off in forested parks. A first glance at the numbers suggests that becoming lost in wilderness is rare. One account of lost-person incidents in a large wilderness park in western Canada records that there are about 26 reports of lost persons each year in the park. Of those 26 reports, about two-thirds of lost people find their own way back to their points of origin, and many of them argue that they were never lost in the first place (though, if my own experience is any guide, there is probably some denial).[2] Given that the total number of visitors to the park annually numbers in the hundreds of thousands, 26 might seem like an extraordinarily small number. Most park visits, though, consist of picnics, short visits for swimming, or the occasional hike down a short trail. If the numbers were to include only visits that consisted of extensive walking through wilderness, especially where such walking took people away from established trails, the proportion of lost-person incidents would be considerably higher.

My own brief experience as a lost person in Algonquin Park pales in comparison with those of the truly lost. The line between a dramatic family anecdote and a life-ending experience is thin and easily crossed. The scientific study of how such victims act, and how to go about finding them, is serious business.

Dwight McCarter, an experienced forest ranger and tracker, has documented accounts of several such cases in the gripping book *Lost! A Ranger's Journal of Search and Rescue.*[3] Some common themes appear in the stories he tells. Otherwise sensible people wander off marked trails for many reasons. They can be looking for a shortcut (as in my own case), confused because of bad weather, or reacting to emotional disturbance (teenage angst or rebellion, or a disagreement with other members of the group). Commonly, losses occur when individual group members have differing abilities and one member lags behind the rest. Once out of sight and earshot, the temptation to try to shortcut, turn back, or do some other unwise thing can be irresistible. Once truly lost, the prognosis becomes grim with astonishing speed, especially when cold temperatures are involved. As time goes by, the likelihood of the lost person engaging in tragically misguided behavior increases quickly.

Many lost individuals are discovered with much of their clothing missing. Hypothermia can produce a burning sensation that the victim tries to ameliorate by disrobing. Many lost persons, especially after lacking human contact for several days, will begin to avoid any searchers deliberately. Children fail to respond to shouts and may even hide from would-be rescuers. Even adults, when searchers make close approaches, may respond with abject terror, presumably because the anxieties that have welled up over the hours and days of isolation spill over into panic at the sound of a rescuer crashing through the woods. Such factors conspire to make what is already a daunting task even more difficult.

Despite the finding that most victims are eventually discovered within a mile or two of their point of disappearance, the amount of ground that is covered by a circle of one or two miles' radius, especially if the terrain is complex, is such that it may take hundreds of searchers many days to complete a comprehensive search.

Elaborate mathematical formulas, based on the nature of the person who has become lost (age, background, reason for being in wilderness) and the type of terrain (changes in altitude, bodies of water), are used to constrain an intensive search to the areas most likely to yield success. Nevertheless, the best chance for survival of a lost person comes at the initial "hasty search," in which a few fit, experienced searchers are dispatched in the area immediately surrounding the point of last contact. If the hasty search fails, the odds against survival steepen dramatically. Unless conditions are very mild, those lost in the wilderness for more than 24 hours are in real peril of losing everything.

WHY ANTS DON'T GET LOST

Rüdiger Wehner has spent most of his professional life wandering the Sahara Desert in search of ants. One ant in particular, the *Cataglyphis fortis,* a bit less than a centimeter in length and weighing about 10 milligrams, has occupied most of Wehner's scientific curiosity for a career spanning many decades.[4] It isn't unusual for scientists to devote staggering amounts of time and attention to seemingly arcane subjects, but there are few cases where this attention has been as richly rewarded as it has been in Wehner's case.

Desert ants are scavengers searching for insects that have succumbed to the rigors of life in a harsh environment. When such victims are found, the ants collect the carcasses and then return to the nest. Though this pattern of behavior sounds simple enough, Wehner noticed something remarkable about the movements of the ant. Ants that are looking for food meander in seemingly random and circuitous paths that can carry them far from their nests (up to about 200 meters). Wehner's surprising discovery was that, on their return to the nest with the food, the ants strike out on a direct, straight-line path for the nest.

Unlike wandering human bush travelers, these ants maintain a seemingly iron grip on their location. How do they do it? One possibility is that the nest emits some kind of signal, such as a smell, that the ants can easily pinpoint. It is well known that ants sometimes follow each other's paths using odor trails, so this seems like an obvious possibility. But using a simple but clever technique, Wehner proved that the ants were not following a scent. When foraging ants reached food sources in the desert, he picked up the ants and moved them to a new location. The ants responded to this displacement by running immediately in a direct course to where the nest *would* have been located if they had not been displaced. This proved that the ants were keeping track of the location of the nest by means of a continually updated estimate of the location and distance to it. The ability to keep such a careful record of one's movements, and to extract an estimate of one's current position from this record, is referred to as path integration. Path integration is one of the chief navigational tools possessed by many different types of animals, such as Wehner's ants, which so excel at it that they can be considered to possess a biological version of Ariadne's thread.

Having shown that ants were making use of path integration rather than a signal that was continuously emitted by the nest, Wehner set out to determine how their knowledge of their own position was established and maintained. There are two separate requirements for the ant: to keep track of the direction of the nest, and to keep track of its own distance from the nest.

We saw earlier that the vestibular system, at least theoretically, can be used to keep track of self-motion. Ants don't have a vestibular system, but they have nicely developed eyes and a sophisticated visual system that is sensitive to a property of light called polarization. Because the human eye is not sensitive to polarization, we cannot know exactly how the world looks to an ant, but it is not hard to understand polarization. Imagine that you have just thrown a stone into a still pond. You will see a series of waves in the shape of concentric circles moving from the place where the stone entered the water. If you were to look closely at one of these waves, you would notice that though the wave is moving outward across the water, the water itself is moving up and down. What makes the shape of the wave is the *vertical* motion of water, but the wave is propagated *outward,* across the surface of the pond.

Exactly the same thing is true of light. Light waves are propagated from objects to our eyes, but there are other aspects of wave motion in light just like the vertical motion of water in the pond. Natural sunlight is said to be unpolarized because all directions of wave motion are mixed up and present in equal proportions in light. The earth's atmosphere acts as a kind of filter for sunlight such that some wave motions are strengthened while others are weakened, producing partially polarized light. How much polarization is present in the light from the sky depends on the position of the sun. If we were able to see the pattern of polarization of light across the sky, we could estimate the position of the sun. (Interestingly enough, this would be true even if the sun were behind the clouds.)

Ants, along with many other insects and a few other animals, can see such patterns. Wehner proved that ants used polarization patterns to find their way home. He designed a cart with a large window fitted with a special filter, much like Polaroid sunglasses, that affected the polarization of light on its way to the eyes of the ant. By following ants across the desert floor with this cart so that the window was always between the sun and the insects, Wehner was able to cause the ants to make errors in homing. In a way, Wehner was using a simple variant of the method pioneered by Tinbergen: the most powerful way to prove that something is being used as a source of information is to manipulate the information in such a way that you can predict errors and then see if the errors conform to prediction.

The compass in the sky provided by patterns of light polarization can tell the ant about the directions of its turns, but it cannot tell it how far it has ranged from home. Ants may solve the distance conundrum in several ways. One idea is that ants measure distance in units of effort. Just like us, ants may know when they've taken a long walk because they feel tired. To test this, Wehner's group trained ants to retrieve food from a feeder while wearing tiny backpacks containing heavy weights (up to four times their own body weight) during the outward-bound parts of foraging expeditions. If effort is measured to estimate distance, then the extra effort required to carry the weight would be expected to produce errors in distance estimation on the (backpack-free) way home. Wehner showed that the weights had no effect. [5]

Another possibility is that ants use optic flow to compute distance. Just as airplane pilots use optic flow to judge their distance from the ground, it might be that ants can measure these fields of flowing movement to calculate how far they have walked. Experiments with honeybees have shown that optic flow can be a powerful source of information about distance traveled, but results with ants have been more equivocal. In one experiment, ants were trained to run along alleyways to obtain food. The alleyways were marked with black and white stripes that would produce nice patterns of optic flow as the ants traveled along them. Having grabbed the food, they were removed from the alleyway and placed on an adjacent alleyway that also contained stripes, but of a different width. When the ants were released in the second alleyway, they would attempt to run back toward their nests. If they were using optic flow, manipulating stripe width should have caused them to make errors. Thinner stripes on the homeward journey should have made them stop running too soon and thicker stripes should have made them run too far.

Although this was exactly the result that Wehner obtained, it was not the end of the story. Ants that had the lower parts of their eyes covered with black paint so that they couldn't see the ground still ran accurately to the nest. So it looks as though optic flow *can* influence perception of distance in ants, but they can calculate distance even when they can't see optic flow. In a way this makes good sense: in their natural setting, running along salt pans in the desert, there would probably be little visual texture on the ground, and so optic flow information might not be prominent.

A third possibility is that ants count their own steps. Wehner tested the "ant odometer" hypothesis by both lengthening ants' legs by gluing tiny stilts to them (made of pig hair, in case you're wondering) and shortening them using, well, scissors! Ants whose leg lengths were altered in this way made predictable errors in nest homing, suggesting that these tiny creatures do, indeed, count their steps to find their way home.

If ants count steps, they do so in intelligent ways. In another experiment, Wehner trained ants to run over a steep hill to reach a source of food. On the homeward trip, the hill was removed. If ants simply counted steps, one would predict that they would run much too far on the homeward leg of the trip, but this is not what happened. Somehow, the ants were able to correct for the change in altitude, arriving safely home after having run just the right distance to the nest. How do the ants keep track of ground distances while traveling over hills? As Rüdiger Wehner says, this is a mystery whose solution "remains to be unraveled."[6]

Before we move on to consider the path-integrating abilities of beings less capable than ants, we should get an idea of the precision of path integration in ants. When the ant completes a foraging run and makes a dash for home, how accurate is it? A typical foraging run carries an ant to a distance of about 200 meters from home. It travels over a meandering course of at least twice that distance before it finds food. At this point, it turns toward home in a path that intersects the outbound path several times but for the most part takes the ant across territory that it has never encountered before. Though the path is not perfectly straight, there is no suggestion that the ant pauses to search for the nest entrance until it is within about 1 meter of home. Using a generous estimate of 1 centimeter for the body length of an ant, this suggests that path integration as used by ants can yield accurate fixes on the nest from a distance of at least 20,000 times its own body length. Translating into human dimensions, using an estimate of 1.8 meters for the height of the average human, an equivalent feat would be to conduct accurate path integration from a distance of about 36 kilometers, a bit less than the length of a marathon. Try to imagine being able to walk in a meandering course, changing directions randomly, for a distance of about 70 kilometers (at a brisk pace this would probably take you about ten hours, not including time to rehydrate and nurse blistered feet), perhaps tackling a few craggy 400-meter peaks along the way. At the end of the walk, without using visible landmarks and without being able to see the point at which you began your walk, imagine being able to turn toward home with a precision of less than 10 degrees. To equal the performance of the ant, you should also be able to estimate the distance to home to within about 200 meters.

As far as we know, no other animal can path integrate as well as the desert ant. Given the lifestyle of these hardy critters, this is not too surprising. These ants live in barren conditions with sparse local cues to help them navigate, and climatic conditions where mistakes could be quickly fatal. To see how our own abilities to path integrate compare with those of other animals, it would be more sensible for us to look somewhat closer to home in the great tree of life. Fortunately, there is no shortage of research on path integration in mammals like us.

One of the first path-integration experiments with mammals was conducted by a husband-and-wife team of psychologists, Horst and Marie-Luise Mittelstaedt. The Mittelstaedts used female Mongolian gerbils for their experiments, and they took advantage of the well-honed maternal instincts of nursing mother gerbils. When young pups stray from the nest, their mothers are diligent about seeking out their errant children. They pick them up gently by the scruff of the neck and return them to their nest. Even in complete darkness, mothers can retrieve their pups by localizing the tiny, high-pitched squeaks the pups produce when separated from them. The Mittelstaedts designed a circular arena with a small container on the outside edge that could hold a nursing mother and her pups. While in complete darkness, one of the pups was removed from the container and placed in the middle of the arena. The mother would instinctively begin hunting for the lost pup. Like a desert ant, the gerbil mother would search in a somewhat meandering path, but once the pup was found the mother made a beeline for the nest, just like the ants collecting food in the desert.

To test whether the gerbils were using path integration, the Mittelstaedts added a small platform to the center of the arena, which could be rotated at different speeds. Again, a pup was removed from the nest and placed on this platform. When the mother stood on the platform with the retrieved pup in her mouth, the experimenters rotated the platform. If they rotated the platform very slowly so that the mother could not sense the movement, she set off for the nest in the wrong direction. The mother's homeward course could be predicted simply by the magnitude of the platform rotation. This proves that, like Wehner's ants, the Mittelstaedts' gerbils were using path integration to track their spatial location relative to the nest.[7]

Similar experiments with other animals have suggested that the ability to path integrate is common in nature.[8] Hamsters led across a large space in the darkness, following a choice morsel of food like the legendary dangled carrot, will turn and run to a hiding place once they have received their treat. Dogs shown a biscuit and then led away on a winding course while wearing a blindfold and headphones can, when released, turn and run to the location of the food with considerable accuracy. Though the path-integration abilities of dogs, gerbils, and hamsters are impressive, there have been no tests of the ability of an animal to path integrate over the same spatial scales as routinely tested in ants. In a way, this would not be a fair comparison. Because of their sensitivity to light polarization, ants have a built-in compass that can always be used to assess direction. Most other animals don't have such an accurate compass and must rely entirely on a record of their own movements obtained from their vestibular system. To understand why this is a disadvantage, we will need to turn our eyes upward.

IT IS ROCKET SCIENCE!

In another desert, far from the home turf of the African desert ant, Robert Goddard toiled in the heat of Roswell, New Mexico, following a boyhood dream to send rockets far into space, a prelude to a mission to Mars. Goddard's quest began at the dawn of the twentieth century when, as a seventeen-year-old boy, he sat in the bough of a cherry tree, looking down at the ground and imagining the view from Mars. He dreamed of a rocket that not only could escape from the earth's atmosphere but could be guided to a target using some kind of navigational system. Later in life, as a rocket scientist in the New Mexican desert, Goddard designed a navigational system based on the gyroscope, a device invented 300 years earlier by the French scientist Léon Foucault.[9]

The wedding of Goddard's solid rocket boosters and Foucault's gyroscope produced one of the great shapers of twentieth-century world politics: the ballistic missile. The navigational problems of ballistic missiles are not very different from those of nursing gerbils finding their way home in the dark. In both cases, knowing where you are means understanding where you've been. In rocket mechanics, such problems are solved using a clever combination of accelerometers and gyroscopes.

A basic accelerometer can be thought of as nothing more than a mass, a spring, and a ruler. As the mass is accelerated, it exerts a force on the spring that causes the spring to stretch. The ruler measures the extent of the stretch, and this measurement yields the size of the acceleration.

As shown in Figure 3, gyroscopes are commonly constructed using a series of rotating rings called gimbals. As the object that carries the gyroscope rotates through space, the gimbals rotate. Measuring the size of the rotation can generate information about changes in heading, or direction.

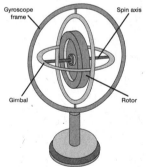

Figure 3: The rotating wings of a gyroscope provide directional information

Both gyroscopes and accelerometers rely on some basic physical laws describing how things that contain mass move. Anything with mass contains inertia, which can be thought of as resistance to movement. Gyroscopes and accelerometers, because they rely on inertia, are said to be the instruments of "inertial navigation." Together, these machines provide all the information necessary to calculate position, provided that the arithmetic can be worked out.

Inertial navigation is very difficult to do well over long periods of time. The path of a vehicle carrying accelerometers and gyroscopes can be reconstructed from the entire record of every *change* in heading or velocity, provided one knows exactly *when* these changes took place. But here's the problem: no inertial guidance system has perfect precision. For that matter, no machine has perfect precision. Every measurement of acceleration or heading change contains an error, and these errors will accumulate as inexorably as the interest on delinquent income tax payments. This kind of inaccuracy, called integration drift, will become more and more serious as time goes by. There are two main ways to counteract this kind of error. One solution is to have a means of measuring velocity that does not depend on the inertial guidance system. Another is to allow the guidance system to come to rest. When the machine carrying the system halts, the velocity falls to zero and so does the value for integration drift. Both these error-correcting mechanisms are used, but the second one is obviously useful only on the surface of the planet, where friction and gravity can bring systems to a halt. It is not very good for rocket navigation, where making things stop can be tricky.

Though the gyroscopes and accelerometers in our middle ears look markedly different from those found in missiles and rockets, the principles involved are exactly the same, and the vestibular system found in mammals suffers from the same type of integration drift.

An animal using its vestibular system for navigation is subject to an accumulating error. Every time the animal turns or moves forward, the error for that movement segment is added to the error from all previous movement segments. Although ants also suffer from this cumulative error, the intrinsic error of their estimates of the sizes of turns is smaller than for mammals because the sun compass can yield more accurate estimates of turn size than can the vestibular system.

Given all this, we would expect path integration based on inertial guidance to be less accurate in mammals than it is in ants, and this is an expectation that has been confirmed in every case so far. But path integration can be carried out with other senses.

When furry creatures try to navigate in the lightless confines of a psychology experiment, they can be made to lose their way like human beings stumbling through a dark house during a power failure. It may take longer for them to stumble, but sooner or later, with enough twists and turns, integration drift will take its toll. A brief flash of light, though, like a bolt of lightning seen through a bedroom window, can reset our sense of position and turn integration drift back to zero. The details of how path integration works in darkness, and how brief visual "fixes" can reverse the accumulating errors of integration drift, have been worked out in experiments carried out with hamsters (their nice habit of stuffing food into cheek pouches and carrying it home to store in a larder makes them an excellent species for studying such problems). The main finding from these studies is that, provided there is not a great discrepancy between where a visual fix tells us we are and the location indicated by our inertial guidance system, a brief glimpse will wipe clean the slate, and re-zero integration drift. If the visual fix gives a surprising result, then it might be ignored.[10]

For example, imagine that you've arrived at your cottage, late at night, and need to find the main power switch to turn on the lights. You have a rough idea of which way to go from the front door, and you set off with your hands in front of you, feeling your way through darkness. A car drives past on a nearby road, and the sweep of headlights through the window provides you with a momentary visual fix. If the fix shows that you are walking at a slight angle to the target, you will correct your course. But if the flash of light suggests that you are walking in the completely wrong direction—back toward the door rather than toward the far wall where the power switch is located—you might be prone to disbelieve your eyes, wondering if the poor light has caused some kind of illusion.

WHAT'S GOOD FOR THE GOOSE MAY NOT BE GOOD FOR US

In a classic study of animal navigation, Ursula von St. Paul took a group of domestic geese on a country ride in a small covered cart. The ride began at their home and proceeded through a series of complicated switchback turns along narrow lanes through varying types of terrain. For some segments of the ride, von St. Paul covered the cart with a blanket so that the geese were not able to see anything. For other segments, the cart was uncovered so that the geese could see the sights as they rode. At the conclusion of the tour, von St. Paul took the geese out of the wagon and released them. Would they be able to find their way home?

The key finding in this experiment was that the geese picked a homeward route as if the only movements they had made had been those undertaken while the cart was uncovered. While carried around in the cart, the geese would have had very little access to inertial guidance because their vestibular systems would not function well while they traveled passively in such an unnatural conveyance. But the most interesting aspect of this finding was that the geese apparently *were* able to path integrate using the flow of visual motion that they received while the cart was uncovered, and this is a very different form of path integration from that using the vestibular senses for inertial guidance.[11]

Although this experiment showed clearly that path integration works well using vision, surprisingly few studies have aimed at determining how precisely this information can be used, especially in mammals. There are a few very intriguing suggestions, though, that geese are not the only animals capable of using path integration in this way.

In many laboratory studies of spatial navigation, animals are carried from their living quarters to another room that contains a testing apparatus. Many researchers have discovered that animals such as rats and mice are actually able to keep track of their orientation to the world as they are conducted passively through the hallways of a large laboratory. Some experiments require that animals lose all sense of spatial connection with the world outside the walls of the testing room, so that the experimenters can be confident that the behavior of the animals is under the complete control of cues present in the room. It can be extraordinarily difficult to produce that state of spatial detachment in an animal. In research in my laboratory (at the University of Waterloo), animals are sometimes conveyed from one place to another inside a light-free container that is rotated on a turntable en route to the testing room. If this is not done, animals often show signs that they have managed to maintain a consistent sense of direction and distance between the room in which they live and the room in which the experiment takes place, presumably by using a combination of optic flow and inertial guidance to keep track of their paths.

Findings such as these suggest that these animals hang on to their sense of place with great tenacity. Draconian disorientation procedures, including incarceration in dark boxes and on spinning turntables, can affect performance in some types of tasks, but even here one of the most peculiar things is that animals, rather than falling back on other sources of information not affected by such procedures (such as landmarks), will sometimes behave as though, having been robbed of their primal attachment to the earth, they cannot make proper sense of these other beacons of navigation.

Do human beings use the same strong sense of direction in solving navigation problems? Edward Atkinson's experience in an Antarctic blizzard, the many anecdotal accounts of people becoming lost in wilderness, and our own tendency to become lost in relatively simple environments like shopping malls and office buildings suggest that we are built differently. Now we should see what the scientific studies have to say.

As a young woman is ushered into a psychology laboratory, she is asked to don a pair of opaque goggles to occlude her vision and a set of headphones to muffle ambient sounds. She communicates with the experimenter via a tiny earbud speaker. The woman is led about the room to one invisible target after another and is allowed to touch each one in turn. At the conclusion of this learning phase, she is led to a starting position and directed to walk to a specified target. Except for the fact that she is in a completely unfamiliar environment, this woman is in similar circumstances to the hapless cottage owner stumbling about looking for the light switch, or the exhausted mother trying to find her way out of a dark bedroom to the sounds of a crying infant. In experimental psychology labs, we can measure with great precision the extent to which people in such circumstances can find their way around, and the results are interesting, to say the least.

In one of the first such studies conducted in my laboratory, we had participants stand at one end of a standard squash court, take a good look around, and then let us lead them from place to place while they wore opaque goggles. We led participants to a series of different objects, one at a time, and then we led them back to the starting position and asked them to strike out by themselves to find the objects. One of the first odd things we noticed was that even though the participants had had a good look at the size and shape of the room, they would raise their hands before them so as to avoid collisions with the walls, even though in many cases the nearest obstacle was at least five meters away from them. Discussions with participants both during and after our procedures soon made it clear that they had very little idea where they were. Our formal measurements of their performance in simple tasks designed to test their knowledge of their own positions revealed that they were performing at levels barely distinguishable from chance. In these initial studies we made no deliberate attempt to disorient them—no lightproof carts or spinning turntables were required for our participants to become completely disoriented. The differences in behavior between people and other animals could not have been more striking.[12]

Through years of experience, we have learned many tricks that have helped us to extract reasonable performance from participants in experiments such as these, but there is still a massive contrast between the performances of non-humans in our laboratory and those of the people who volunteer for our studies. With animals, the challenge is usually to find a way to make them forget about the larger spatial context of the labyrinth of laboratory rooms so that we can be sure we're controlling how our critters use space. With people, the challenge is to provide them with enough support that they can find their way across an ordinary rectangular room without banging their heads into the walls. Why do such differences exist? Though there's still much that we don't understand about this, there are some important clues.

In very simple situations, we can find our way to a target with reasonable accuracy. For example, imagine a task in which you are able to take a long look at a target that is lying on the ground some distance in front of you—say about 10 meters. Then, with eyes closed, you are asked to walk to the target. Provided that you are allowed to walk immediately after you close your eyes, you should be able to land within a few centimeters of the target. As you read these words, you may be very skeptical that you could perform well in such a "blindwalking" task. When you have a chance, try it out (ideally in a large, flat, outdoor space like a sports field). You will almost certainly be surprised by your accuracy.

When walking tasks like these are made slightly more complex, human performance unravels quickly. On a triangle-completion task, blindfolded people are walked along a path of a few meters and then, after changing direction, they are led along a second path. Their task is to complete the triangle by walking back to their starting point. There are two main differences between the triangle-completion task and the blindwalking task, both probably important. First, the triangle task does not contain an explicit visual preview of the target or of the stopping points (the corners of the triangle). We need to plan our homing route entirely on the basis of bodily information that we receive while walking (vestibular information, feedback from the muscles used in walking, and so on). Second, the triangle-completion task involves measuring both a pair of walked distances and the angle between them—considerably more complex than just estimating a visual distance. These two differences conspire to degrade our performance on this task sufficiently that, as ants, we would surely starve or fry in the blazing sun. In one typical study, after walking short triangles ranging in size from two to six meters, the average angular turning error in heading for home was more than 20 degrees, and the size of the distance error was around 50 percent of the distance walked.[13]

Except when the power fails or when we're trying to creep around darkened rooms without disturbing sleeping family members, we human beings are not very likely to find ourselves trying to navigate entirely without using our visual sense. But the fact that we become disoriented so quickly and completely when deprived of visual fixes has a greater meaning. Our inability to tap into body-based senses to keep track of location may be a deficit rooted in our biology, a loss of an ability possessed by our ancestors that has fallen into dormancy through lack of use, or, what I think most likely, a combination of the two.

If we once possessed the ability to keep track of our location using path integration, perhaps even if not to the same degree seen in ants or even in rats or geese, what has caused our increasing tendency to lose contact with position, place, and space? Has some other way of understanding space come to supplant the ancient ways of gluing us to our place on the planet? The landmarks are now in place, our route is becoming slightly clearer, and some of the answers lie directly on our path.

CHAPTER 4
MAPS IN THE WORLD

How Expert Navigators Use Specialized Senses to Find Their Way

Every cubic inch of space is a miracle.
WALT WHITMAN

Navigation using a map is a key transition from the simple kinds of tasks that I've been discussing thus far to the more complicated accomplishment of true wayfinding. In a wayfinding task, not only is the target invisible from the starting point but it can be found only by carrying out the correct sequence of movements based on what can be seen, heard, or felt at each point in the sequence. You don't *need* a map to complete a wayfinding task, and you *certainly* don't need a map for any task that is simpler than the definition I've just given.

When you read the word *map,* the most likely thing that will spring to your mind (unless you're a mathematician) is the folded paper that you might find in your glove compartment. There is no question that this is a map, albeit a very specialized one. The reason the map is taking up space in your car is that it is a useful tool for navigation, and what makes it useful is that the map is a model of the real world. A good map will contain replicas of things found in the real world that it is useful to know about, such as roads, schools, and shopping malls. Maps often have other useful features as well, such as a compass rose that allows you to orient the map properly to the real world and a scale so that you can work out the real distances between the points represented. But do all of these features have to be present for something to be called a map? To answer this question, we'll want to stand back and take a much more general view of maps, where they come from, and how they are used.

I have a friend who has spent much of his life traveling from one country to another in search of many of life's greatest pleasures, including fine beers. Because the consumption of fine beer is often incompatible with razor-sharp accuracy in calculations of currency conversions, my friend has learned a useful trick. He carries a card in his wallet that lists a series of dollar amounts and their equivalents in local currency. To a mathematician, this relationship of one set of values to another constitutes a map just as surely as the one in the glove compartment. The main difference between this kind of map and a road map is really only that the road map contains two dimensions whereas the currency chart contains one. Direction isn't part of a one-dimensional map, so a compass is not required. In both cases, a scaling factor is used to relate one variable to another.

Mathematicians who are interested in different ways of mapping one quantity on to another are called topologists. The formal definition of topology is almost guaranteed to cause you to put this book down and run away screaming, so I will give only an informal idea of what topologists do.[1] Imagine taking a sheet of something flexible, like rubber or latex, and drawing a simple map of your neighborhood on it. Now think of all of the ways that you could distort that sheet by, for example, stretching it over your face or stepping on one side and tugging at a corner. The only things that are against the rules are ripping the sheet or gluing any of its edges together. A topologist is interested in understanding which properties are preserved by your wanton handling of this map and which ones are altered.

For mathematicians, the importance of topology is hard to overestimate. Not only does it draw links between major areas of mathematics such as algebra, geometry, and mathematical analysis but it has also led to the mathematical field called graph theory, which has been pivotal in providing the tools to help provide solutions to such practical matters as how to prevent traffic jams and how to design networks of computers. Many problems in applied mathematics involve finding the most direct and efficient routes between one place and another. One classic example of this sort is the "traveling salesman problem," in which one has to find the most efficient route that provides one visit each to a group of randomly arranged targets. The traveling salesman problem is of interest not only to, well, salesmen but also to those who design such things as circuit boards (to minimize production costs) and robotic devices that carry out repetitive tasks.

In psychology, the field of topology has helped us to understand the ways in which maps can be used to navigate. For example, think of the last time you drew a sketch map for someone to help them find their way from one place to another. Typically, such maps contain only the bare minimum of features that you deem necessary for them to find their way without becoming lost, so the major emphasis is placed on those points where people need to make explicit decisions ("Which way do I turn when I see the post office?"). Scale is seldom well preserved on sketch maps. A long, straight stretch of highway is likely to be compressed compared with another part of the map where a complicated cluster of intersections needs to be negotiated. So your map is a good topological map in the sense that it preserves the connectedness of places and the order in which they will be encountered, but it is not what is called a metric map because dimension and angle have not been preserved. For the most part, we expect the paper maps that we buy at gas stations to be metric in this sense. We expect there to be a proper scale to them, so that they represent accurate two-dimensional models of the real world.

In a later chapter, we'll revisit the artifactual maps—the kinds of representations of space that have been drawn by human beings for thousands of years—to help us to understand how humans mentally represent space. Right now, though, we want to step back and take a somewhat broader view of the different types of maps that are used by both animals and human beings to guide themselves through space. As with some of the earlier parts of our story, some key differences between humans and animals will emerge from our exploratory forays through the animal world.

HOW HOMING PIGEONS FIND THEIR WAY

Racing, or homing, pigeons have a distinguished history as long-distance navigators. These birds, the same species as the common city-dwelling pigeons, have been used to assist with human communication for thousands of years. Before the widespread adoption of the telegraph in the nineteenth century, sending messages by pigeon was one of the fastest ways to communicate over long distances. Ancient Egyptians may even have used pigeons to announce the coronation of Ramses II. In a tradition perhaps beginning with Genghis Khan, Julius Caesar is said to have used pigeons extensively to convey battle knowledge, especially in his conquest of Gaul. The Reuters news service began as a system by which financial information could be conveyed from Germany to Belgium via telegraph wires, but the gaps were filled using what some pundits have recently referred to as the wi-fly method—pigeon telegraph. For as long as humans have used pigeons to carry bits of paper over great distances, we've also been interested in the mechanisms that they use to find their way reliably from one place to another. The scientific history of our interest in pigeons begins with Pliny the Elder in the first century A.D. In his astonishing encyclopedia, *Naturalis historia,* Pliny reports that Decimus Brutus used pigeons to convey news of the outcome of a battle at Modena.[2]

Consider this scenario. A homing pigeon is removed from its loft and placed inside an opaque box, which is fastened to the back of a truck, and transported over a sinuous route for a distance of 100 kilometers. The box is opened and the pigeon flies away. The initial flight direction of the pigeon seems random, but within a short time the bird is flying in a direct path back toward the loft. In a matter of hours, it has returned home. Such acts of successful homing can take place over distances of thousands of kilometers. As with so many of the navigational feats of animals in long-distance migration and homing, we know less than we'd like to about how such prodigious accomplishments come about, but it seems clear enough that in order to carry out such journeys, pigeons need two things: they need to know where they are and they need to know where they are going.[3]

Many useful sources of directional information exist in nature. Sun, moon, and stars, provided one knows something about the passage of time, can be used as a compass. Our own pocket compasses rely on something different—the earth's magnetic field. The magnetic field of the earth is often portrayed as something like a huge bar magnet thrust through the center of the planet, with one pole roughly corresponding to each of the planetary poles. Though this is a handy diagrammatic shorthand for understanding how a compass works, it would be a mistake to believe that the physical processes that give rise to the earth's magnetic field share much with a bar magnet. The real cause of the field is something that geophysicists call the dynamo effect, which is caused by the movements of massive amounts of conductive molten iron deep within the planet's core. These movements, caused in turn by the rotation of the earth, throw gigantic magnetic field lines across the surface of the planet and far out into the space surrounding it. When we hold a small navigational compass in our hand and watch the needle align with magnetic north, we are witnessing an alignment between the slender rod of metal in our hand and these huge churning seas of molten rock and metal deep beneath us.

Do pigeons use a magnetic compass to find direction? An animated battle rages among scientists who study pigeons, but those who believe that pigeons can use magnetic fields to navigate are slowly gaining ground. Some early studies showed that forcing pigeons to fly while wearing tiny coils that disrupted the magnetic fields surrounding their heads caused them to become lost. Though this would seem to be incontrovertible evidence for some kind of use of magnetic fields to navigate, the effect was largely confined to young, inexperienced birds. The old hands were much less disrupted by such treatment, for reasons that are now becoming clear and which we will look at in a moment. The biggest obstacle to establishing the use of magnetic fields for compass orientation in birds has been the difficulty of finding the magnetic receptor. Sounds are detected by ears, smells by nostrils, and sights by eyes. What sense organ detects magnetic fields? It turns out that the answer to this question might have eluded us for so long because the receptor organ was, so to speak, staring us in the face the whole time. Though the fine details remain to be worked out, it appears that pigeon photoreceptors, the cells in the back of the eye that convert light into electrical impulses, are sensitive to magnetic field orientation. Just as ant eyes can detect polarization angles, pigeon eyes seem to be able to detect magnetic field properties. We have no way of knowing whether pigeons can actually "see" magnetic fields, but there is at least good evidence that the direction of these fields can influence the way pigeon photoreceptors work. Regardless of whether these influences rise to pigeon consciousness (whatever that might be), they might be enough to influence flight direction.

The evidence for a magnetic sense in pigeons helps explain how they complete their long journeys, but one major piece of the puzzle continues to elude us. In the example that I described earlier, in which a pigeon was released from a box into an unfamiliar neighborhood, it was able to find its way back to its loft. Because the pigeon had never been to the release area before, it wouldn't be enough for it to know which direction was which. It would also need to know something about where it was. Without a map, our pigeon is still lost.

Though pigeons may be the most intensively studied case of this type, some other animals share the pigeon's ability to find home from an unknown location, suggesting that they too can map their own location on the planet's surface, even when they are released in an entirely new location. Current thinking suggests that the only way to accomplish a feat like this is to have what is called a gradient map.

Imagine that you are standing in a large, square field. On one side of the field a noisy road crew is doing some repairs with a pneumatic drill. On an adjacent side of the field a street vendor with a food cart is playing a loud, repetitive jingle. With your eyes closed, you could wander around in the field and work out your distance from either the road crew or the food cart by gauging the loudness of the sounds. Knowing both distances would allow you to triangulate your position on the field with an accuracy limited only by your ability to discriminate loudness. What is even more interesting about this example is that you could work out your position in the field even from locations that you had never visited before, provided you had a basic understanding of the principle—two sources of sound in two different locations provide unambiguous cues to position.

Maps based on these principles work by taking advantage of some kind of gradient—some feature of an environment that changes in a regular manner depending on the observer's position. Sounds are louder when we are closer to them. Smells are more intense. Lights are brighter. What about magnetic fields? It turns out that these fields vary systematically across the surface of the planet as well.

Indeed, the properties of magnetic fields that are used by navigating animals are more complicated than the properties that are used to help us point ourselves toward north in a cluttered forest. Geomagnetic fields have both intensity and direction. So for any point on the ground, imagine that there is a set of invisible arrows representing the geomagnetic field in that area. The arrows not only point in a particular direction but also have a specific length (which corresponds to intensity). Both direction and length vary systematically according to one's position. From the standpoint of the gradient map principle, one of the most useful things about this is that the two properties of the geomagnetic field are somewhat independent of one another. Because field direction can change independently of field intensity, knowing both of these values can uniquely define one's position in space. In addition, there are lots of interesting irregularities in the gradient map formed by the earth's magnetic field. These irregularities are most commonly caused by gigantic pieces of rock with their own magnetic properties that interfere with magnetic field lines from the earth's iron core. The beauty of these features is that they can help to further define the magnetic environment of the pigeon's neighborhood and allow it to localize itself more precisely in space.

So much for theory. Do pigeons actually use magnetic gradient maps? Some convincing evidence suggests that they do. For one thing, experienced birds become disoriented when their ability to read magnetic maps is disturbed either by the presence of anomalous sources of magnetism or by interference with the system thought to be involved in magnetic map reading. In one of the most interesting of recent studies, homing pigeons were released in the vicinity of the Auckland Junction Magnetic Anomaly, an area in New Zealand with an extraordinarily high and unusual spike in magnetic field intensity and direction. Pigeons released here showed wildly disordered flight paths that seemed to be under the control of the local anomaly, but as their distance from the geomagnetic spike increased, their flight directions became more ordered and they began to fly in the right direction for home.[4] Interestingly, the detector system for reading magnetic maps may be entirely separate from the magnetic compass found in the pigeon's retina. Nerve fibers embedded in the nostrils of pigeons and other birds contain specialized magnetite molecules—much like those thought to have been discovered in Martian bacteria—and it is these nerve fibers that appear to be involved in map reading.

It isn't very likely that homing pigeons rely entirely on magnetic fields to find their way around. Although the use of a gradient map might be the only way to account for navigation over vast distances from unknown origins to familiar home sites, some of the more mundane "everyday" navigational problems faced by pigeons could be solved by other means. Like many other animals, pigeons use visual landmarks (including human-built artifacts such as systems of highways), and some intriguing evidence suggests that they might be able to navigate over considerable distances using their finely tuned sense of smell. It has even been proposed that pigeons are able to construct a kind of mosaic map of space based on olfactory panoramas in much the same way as they can use magnetic fields.[5]

Homing pigeons are far from being the only animals that can use magnetic fields to navigate based on gradient maps. Many other birds, as well as aquatic creatures, have a keen magnetic sense. Even some mammals, such as the naked mole rat, an unusually social rodent living a subterranean life in conditions where landmarks are sparse, have been shown to rely on magnetic fields to navigate.

Other than pigeons, one of the best-studied examples of navigation by magnetic field is the sea turtle. Green turtles live and forage near the coast of Brazil, but a large number of them lay their eggs on the beaches of Ascension Island, in the south Atlantic. This small island, conveniently located about halfway between South America and the coast of Africa, has long been used as a stopping-over point for mariners and a staging area for military actions ranging from government raids on slaving ships in the nineteenth century to the Falkland Islands conflict in 1982. Green turtles had the misfortune to become a handy source of fresh protein for sailors, who at one time engaged in the barbaric practice of capturing these huge animals and tying them upside down to the decks of their ships as a kind of living larder. The turtles were hardy enough to survive under these conditions for weeks, waiting to be turned into soup for the ship's crew. Because green turtle soup was also a prized delicacy in Europe, turtles were occasionally captured with the objective of returning them alive to the home port and delivering them with much fanfare to the wealthy and the titled. It was common for such captured turtles to be branded or otherwise labeled with the name of the anticipated recipient. The illness and death of the Duke of Wellington—a turtle destined for the royal table—was a noteworthy enough event to be recorded in a ship's log.

In 1865, Carl Cornelius described one of the most extraordinary features of green sea turtles when he recounted an episode in which such a turtle had been captured near Ascension, branded, and then released in the English Channel when it appeared to be unwell. Two years later, this same turtle was recaptured at Ascension, suggesting that it had found its way through thousands of kilometers of entirely unfamiliar waters from the English Channel back to its foraging ground near Brazil and from there back to Ascension Island.[6] Though this was one of the first reports of the impressive navigational abilities of sea turtles, it is now well known that these animals enjoy a life cycle requiring them to carry out several long-distance migrations. It is virtually impossible to imagine these being completed unless turtles possess a gradient map.

Loggerhead turtles lay their eggs in the sand on the eastern shore of Florida, and the young, following subtle contours of light and terrain, find their way to water. Assuming they avoid being eaten by shorebirds that lie in wait for the hatchling turtles to make their dash to the sea, loggerheads find deep water by sensing the directions of wave patterns and, once free of the beach, rely on magnetic fields to guide them across the ocean to the Atlantic's eastern shore. Like many other ultra-long-distance navigators, loggerheads probably rely on simple combinations of cues, such as the direction of major ocean currents and of magnetic field lines to carry them to favored waters. Though not impossible, it is unlikely that these turtles use a gradient map for the initial transatlantic voyage because the scenery would be entirely new to them. But in addition to these long-haul voyages, loggerheads are capable of surprising finesse in localizing their breeding grounds on the Florida beaches near Melbourne. Displacement experiments in which loggerheads are picked from the sea and returned to it far from their favored breeding grounds have shown that loggerheads' well-developed ability to find their way back to traditional nesting sites seems to rely on the magnetic sense. This sense probably includes an ability to read local microvariations in magnetic fields caused by the peculiarities of rock formations underlying the ocean floor.[7]

DO HUMAN BEINGS POSSESS A MAGNETIC SENSE?

Many years ago, I invited my parents to visit me at a house that I had just purchased. I gave what I thought was a foolproof set of directions from their home in St. Catharines, Ontario, to my house, about a two-hour drive. En route, my father telephoned to tell me they were lost. I had him explain to me where they were, and then I gave him a new set of directions. A few minutes later, the telephone rang again. They were lost again, so more directions were offered. I knew that they were now very close to my house, so I went outside, sat at the edge of the road with a cup of coffee in my hand, and did my best to behave as a highly visible and familiar navigational beacon for them. I was pleased to see their car appear at the end of the street, but my pleasure turned to dismay as I watched them drive right past me in spite of my standing at the very edge of the road, waving both arms madly. Both parents were oblivious to my presence and obviously in the middle of a heated discussion. Like so many married couples, it seems, there were strong differences of opinion about which way they were going, and who was to blame for their wayfinding struggles.

We've all been there. Unsure of our bearings, we pull the car up to an intersection and peer right and left. We make a choice based on ephemeral feelings, gut instincts, and perhaps the quality (or volume) of arguments from our traveling companions. Some of us have a better sense of direction than others—my wife's is better than mine—but we don't know why.

Good friends of ours take advantage of their familiarity with their own senses of direction to find their way. Ian, an engineer, asks Brandy, an adventurous entrepreneur, which way they should go. She replies boldly. He applies what he calls the Leverette Transform (Leverette is her maiden name), which means that he always goes in the opposite direction to the one that she specifies. Though he has had to endure the occasional frosty silence on the drive, he says the strategy works much more than would be predicted by chance.

When we are lost and all navigational cues have abandoned us, we guess. Some of us guess well, and some others not, and little is understood about why this might be. Is it possible that, like pigeons, turtles, fish, and mole rats, we possess an intrinsic sense of direction based on sensitivity to magnetic fields?

In the 1980s, a British researcher named Robin Baker set out to answer this question in a long series of experiments that involved taking human observers on long bus rides into the countryside. The routes were chosen so as to be as baffling as possible, including lots of tortuous twists and turns and preventive measures to make it difficult for participants to use any of their conventional senses to keep track of their position. At the end of the ride, the participants were asked to point in the direction of home. Not only did Baker show that his participants could often do a good job of pointing homeward but he also showed that this ability could be disturbed by asking them to wear bar magnets on their heads or special helmets that produced magnetic fields.[8]

In studies conducted in the laboratory, Baker had participants sit in rotating chairs while being deprived of both light and sound. After spinning them extensively, Baker asked them to point toward north. His data showed some evidence that people could actually perform such a task, but that when they wore magnets on their heads, their responses fell to chance. Baker's studies attracted a flurry of interest at the time, and several researchers attempted to replicate his findings, but with less success than reported by Baker. In spite of these failures, Baker continued to argue vehemently for the presence of a magnetic sense in human beings.

Today, there appear to be few supporters of Baker's conjecture. Though one or two intriguing findings suggest that human beings may possess small deposits of iron in the bone lining the nasal cavity and that patterns of brain activity may be changed in subtle ways by exposure to magnetic fields, no new and unequivocal studies show a magnetic sense in human beings.[9]

Although the scientific effort to identify a human magnetic sense lies fallow, the possibility of such ability is still raised occasionally as an explanation for anecdotal accounts of extraordinary single acts of navigation. Nainoa Thompson, one of the founding members of the Polynesian Voyaging Society, has devoted his life to the effort to preserve the traditional wayfinding methods of Polynesian seafarers. One of his most impressive adventures has been to repeat the 4,200-kilometer voyage from Hawaii to Tahiti in an outrigger canoe using only traditional methods along with some knowledge of modern astronomy. In an account of one of these voyages, he describes an episode in which, late at night, under heavy cloud cover in the notorious Doldrums with their constantly shifting wind patterns, he felt himself to be completely lost, yet moving at a quick clip of about 25 knots—the worst combination of circumstances for a navigator. In describing what happened next, Thompson says that he leaned back against a railing, relaxed, and was overcome by a feeling of warmth and a confidence that he knew the direction of the moon. Acting on this confidence, Thompson found his way.

Ben Finney, a professor emeritus of anthropology at the University of Hawaii, also a seasoned navigator and founder of the Polynesian Voyaging Society, has suggested Thompson's case and a few other similar ones might be explained by a deep, latent magnetic sense that can be tapped by those with vast amounts of navigational experience.[10] Though such arguments by themselves are not enough to turn the tide of scientific opinion, they suggest that a few human beings with the right combination of intuition, experience, and perhaps desperation can tap into resources that are not available to everyday navigators such as quarreling couples driving through unfamiliar towns looking for antique stores and teahouses.

Yet this mysterious capacity to tap into resources that seem completely foreign to the rest of us to produce gradient maps may have less to do with exotic sense organs and more to do with exquisite sensitivities among the more commonly understood sensory systems of sight, sound, and feeling.

———

On my way to a temporary posting in Australia some years ago, I took advantage of a generous travel allowance to make a brief stop in Tahiti. To experience a little bit of Polynesian life off the beaten track, I booked a ticket on a small and ancient ferry that transported people across 18 kilometers of open sea to Moorea, a gorgeous island reminiscent of the mythical Bali Hai that James Michener described in *Tales of the South Pacific*. As I looked down at the dock from my perch on the old rust bucket, I noticed a group of men looking up at me and the other passengers and gleefully throwing small packets of money down to the ground. After our short voyage began, and the boat started to toss about in the rough seas, it occurred to me that the men may have been placing bets on whether any of us would be seen again. I'm not known for my sea legs, but when the man who was seated beside me, an experienced airline pilot, made his third lurching visit to the rail, I realized that I was in the grip of something exceptional. Much to the chagrin of my traveling companion, who seemed to have a constitution of steel and who had hoped for a romantic tête-à-tête en route, my stomach contents and I survived the trip intact only because I found a small spot of dirt on the floor to fixate upon and refused to move my head or eyes for the entire duration of the journey.

It doesn't take much time aboard a small vessel at sea to learn that one will be tossed about by ocean currents, swells, and waves in all but the most unusually calm situations. Though a casual observer will only feel buffeted by random and chaotic movements of water, a keen navigator can use these movements to establish both direction and position.

On the open sea, swell patterns are set up by interactions between directions of prevailing winds and ocean currents. Expert navigators, relying on a detailed knowledge of both of these forces, can read the swell pattern to determine direction. When several swell types intersect, navigators can also use interference patterns to obtain some positional information: the presence of intersecting waves moving in two different directions sets up the minimal conditions required for a gradient map. Under the tutelage of traditional navigators from the island of Puluwat in the South Pacific, anthropologist David Lewis was able to learn to read some of these patterns himself. Suggesting to his teachers in a tongue-in-cheek manner that navigating by swell pattern was real "seat of the pants" work, he was told that navigators learn that the pattern of swells is easiest to detect by paying attention to the pattern of stimulation of the testicles, as, while seated, this was the part of the body that was most sensitive to the smallest movements of the boat.[11] If true, this argument presents incontrovertible evidence for one kind of sex difference in navigational skills!

In open water, swell pattern can indicate direction and, in some cases, position. Near landmasses, complicated patterns of refraction or reflection are established. Evidently, Puluwatese navigators are able to interpret these patterns. The virtuoso navigational performance of the renowned Tevake, described by Lewis in chapter 2, was conducted under overcast skies, so there was no opportunity to use any kind of astronomical information. Under these conditions, the most likely explanation for Tevake's success would be that he was able to analyze swell patterns to compute his position in a manner very similar to that of sea turtles that may use microvariations in the patterns of magnetic fields on the sea floor.

HUMAN TRAVELERS BUILD GRADIENT MAPS WITH SNOW, SAND, AND STORY

In the Arctic, reliable landmarks or inukshuks that can be used to guide trekkers on the tundra are not always around. In the huge central Arctic Barrens, a traveler is more likely to be confronted with nothing more than flat fields of white ground that extend as far as the eye can see in all directions. In this sense, Barrens navigators are in a situation not very different from that of Polynesian seafarers. Though Inuit navigators can use some simple celestial information such as the position of the sun or moon for navigation, they rely less commonly on sophisticated star maps than their more southerly counterparts. One reason for this is that much travel takes place in the summer months, when the periods of darkness can be very short or even nonexistent.

Fortunately, the Inuit have some other resources to help them find their way. Inuit navigators have an understanding of wind patterns that equals the understanding of ocean swell patterns in marine navigators. Even when the winds are not active, their history is etched into patterns in the snow known as sastrugi. Though it would be unusual for patterns of sastrugi or wind to set up the right conditions for a gradient map (because this would require at least two directions of prevailing winds), sastrugi can be used effectively as a kind of compass.

Compass headings given by wind or snow are accurate only to within about 10 degrees, so other tools are required to zero in precisely on targets. Barrens Inuit make use of tracking for these purposes. Long before you reach a target in the Arctic, you are likely to encounter the trails of other parties of dogsleds. Skilled navigators can read these trails in order to determine the identities of the sleds that have passed (based on the width of the runners and even the footprints of individual sled dogs). This information can be used to help achieve more precise navigation. In traditional Inuit societies, social networks are inextricably tied to places, so the facts of one may be used to predict the facts of the other.

In another climate entirely, the Bedouin, a nomadic desert culture, are renowned for their feats of wayfinding and tracking. Donald Cole, a literary traveler, described having experienced these abilities at first hand during an adventure that involved crossing the Great Empty Quarter, a massive expanse of desert in the Middle East, with a group of Bedouin. He described an oft-repeated episode in which he was left in charge of steering a camel while the rest of his group rested. One of his companions would awaken, glance briefly at their surroundings, and then, with a sure and gentle hand, correct Cole's course. In addition to a detailed mental inventory of landmark locations similar to those possessed by Inuit and Aboriginal navigators, Bedouin navigation relies upon extensive knowledge of the positions of the stars. Indeed, many stars were named originally by Middle Eastern observers keen to make sense of the night sky. Even backyard astronomers will have noted the large number of stars identified by their Arabic names.[12]

If anything, the tracking skills of the Bedouin exceed their wayfinding abilities. In the past, Bedouin trackers were routinely attached to police departments in villages in Saudi Arabia. In his classic travel book *Arabian Sands,* adventurer Wilfred Thesiger describes an experience that illustrates nicely Bedouin tracking abilities:

A few days later we passed some tracks. I was not even certain that they were made by camels, for they were much blurred by the wind. Sultan turned to a grey-bearded man who was noted as a tracker and asked him whose tracks these were, and the man turned aside and followed them for a short distance. He then jumped off his camel, looked at the tracks where they crossed some hard ground, broke some camel-droppings between his fingers and rode back to join us. Sultan asked, "Who were they?" and the man answered, "They were Awamir. There are six of them. They have raided the Junuba on the southern coast and taken three of their camels. They have come here from Sahma and watered at Mughshin. They passed here ten days ago." We had seen no Arabs for seventeen days and we saw none for a further twenty-seven. On our return we met some Bait Kathir near Jabal Qarra and, when we exchanged our news, they told us that six Awamir had raided the Junuba, killed three of them, and taken three of their camels. The only thing we did not already know was that they had killed anyone.[13]

In this example, much of the information was carried in the patterns of camel tracks. A good tracker can determine the origin of the camel that made the tracks based on the size, shape, and depth of the tracks, the recency with which the camel fed and drank based on its stool, and the terrain over which it has traveled by the state of its hooves. This knowledge, combined with an intimate understanding of the surrounding terrain and an understanding of local politics, will allow a tracker to put together a detailed story from a few marks in the sand. In other circumstances, trackers have been known to build similar stories by examining human footprints. It is said that excellent trackers can identify a person based on tracks they have made wearing shoes—even if the shoes belong to another person!

Though the Arctic and the deserts of Arabia are vastly different places, the people who live there show similarities in navigational legerdemain. These similarities include heightened attention to subtle environmental details, a tendency to weave these details into the mythology and spiritual life of the culture using stories, and an ability to use inferences about the social life and local politics of other roving groups to judge position and movement, even when available information is sketchy.

For any animal, the holy grail of navigation is the possession of an accurate map. When animals have specialized sensory systems, such maps can be effectively written into the terrain in the form of gradient maps. It is almost as if an animal that can read gradients can measure its current coordinates, compare them with the coordinates of the goal location, and plot a course. Lacking such specializations, we humans can sometimes turn what sensory resources we possess to novel uses, as in the analysis of swell patterns and sastrugi to locate ourselves in space. When even these rudimentary forms of gradient mapping fail us, those whose lives depend on an ability to maintain their sense of place at all times fall back on image, memory, and story to bind themselves to the landscape. Some evidence from the planet's most accomplished human navigators suggests that, when pushed to the outer limits of their abilities, they may not even *know* how they are solving difficult wayfinding problems.

The illustrations of the wayfinding feats of the Puluwat, the Inuit, Bedouin trackers, and others as well suggest that when our ability to know where we are in a natural setting becomes a matter of life and death, we may have access to subtle signals in the environment much like those used by homing pigeons or turtles. Although finding such signals is certainly possible in some cases, it is a skill that has been almost completely lost to modern human beings. One cost of such a loss is the fact that, unless heavily supported by signs, guide ropes, trails, and roadways in our environment, we become lost. We have already seen some examples of such disorientation, and more are on the winding path ahead of us. But a deeper cost to such losses will also emerge as the story unfolds. Losing such wayfinding skills has helped propel a dangerous trend whereby our connections to our natural environment have become seriously severed. The breaking of such connections has consequences for everything from our attachment to nature to the kinds of houses and cities we live in and how they make us feel.

When solving everyday wayfinding problems, we often feel that we are consulting a kind of mental map. Everyone is familiar with this idea. When our favorite route home from work is blocked by road construction, we can quickly conjure an alternative route. When a hapless tourist stops us on the street to ask directions, our eyes may turn skyward for a moment as we whip up a bird's-eye view, a cerebral streetscape that we can use as a tool to guide the stranger to the area's best breakfast or a chic gallery on the other side of town. It seems incontrovertible that minds possess maps, yet the shapes and forms of these maps in different types of animals, and especially in human beings, hold some surprises.

CHAPTER 5
MAPS IN MOUSE MINDS

THE MENTAL MAPS OF SPACE POSSESSED BY ANIMALS

*If we use excessively elaborate apparatus to examine simple
natural phenomena, nature herself may escape us.*

KARL VON FRISCH

Edward Tolman appeared on the scene in experimental psychology in 1918, when he took up a post at the University of
California at Berkeley that he kept for his entire career.[1] The experiments he conducted using rats served as the beginning point of a
modern cottage industry in using rats to understand how minds
deal with space. Tolman began his work at a time when psychology in North America lay heavily in the shadow of a radical theory
that seemed to deny the existence of mind itself. John Watson, the
father of the behaviorist movement, had published a manifesto
in which he claimed that all human behavior could be accounted
for as sets of learned associations between stimuli and responses.
Like Pavlov's famous dogs, we were doomed to go through life

responding more or less automatically to tolling bells and flashing lights, doing only what we had learned brought pleasure and avoided pain.

Though it seems incredibly brutal by today's ethical standards in psychology, one of Watson's most influential papers demonstrated that it was possible to train a young child to fear a white rat using simple conditioning methods. Watson showed the rat to the child while making a loud and startling sound immediately behind the child's head. Not surprisingly, the child quickly learned to fear the rat, and Watson made the bold claim that all human phobias had a similar genesis.

Tolman, in contrast, was reluctant to cast away the idea that human heads contained things called minds, whose composition was much more interesting than the collection of associations envisioned by psychologists like Watson. Not only did Tolman take the then radical view that human beings had minds but he wondered whether his collection of rats might possess such things as well. To test his idea, he devised laboratory tasks to assess how much his rats knew about space.

Rats can be trained to do tasks with great ease using methods much like those you might employ to train a pet to perform a trick. In the beginning, they are given a simple task to perform (leave the start box) and are rewarded with food. As they become more accomplished, they are given greater challenges until, eventually, they can complete the whole task. To begin, Tolman placed his rats in a small, square box and trained them to cross a round chamber to enter a narrow alleyway. After a short run in the alleyway, the rats were required to make three right-angle turns into a final alleyway, at the end of which they would find a tasty food reward. The basic setup is illustrated in Figure 4.

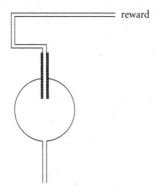

Figure 4: Tolman's special orientation maze for rats

Once his rats had learned their way to the reward, Tolman made a critical change in the shape of the maze. First, he closed off the alleyway leading out of the central chamber partway along, and next he opened up a large number of alleyways radiating from the chamber in the shape of a starburst, as shown in Figure 5. The location of the reward remained unchanged.

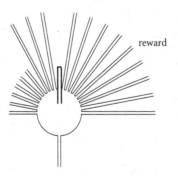

Figure 5: Variation on Tolman's original maze

What will the rat do? According to a behaviorist, Tolman has broken the chain of simple behaviors that the rat has learned. When it crosses the central chamber, enters the usual alleyway, and finds its way blocked, either it will bash into the closed door in frustration or it will wander aimlessly with no clear plan. But what Tolman observed was something much more interesting. Most of his rats chose the alleyway that led most directly to the reward—which, remember, hadn't moved. What this clever response suggested was that Tolman's rats possessed an understanding of the spatial relationship between the start box and the food reward that allowed them to follow a route they had never seen before. Indeed, before Tolman made the changes, the route simply did not exist. It was as if the rats had stored an overhead view of the scene in their minds that they could consult to find the best way to respond to an unexpected turn of events.[2] For Tolman, the discovery of this "cognitive map" was a major victory in his attempt to demonstrate that simple associations between stimulus and response did not adequately to explain all behavior—even the behavior of a laboratory rat. For our story, the main importance of this finding was its implication that rats must possess something like a map in their minds that they can consult to solve the problem.

How do such cognitive maps differ from the gradient maps that guide the movements of homing pigeons and turtles? The main difference is that gradient maps are based on some kind of relatively straightforward physical quantity that is continuously available to animals and can be directly related to position—geomagnetic force lines, patterns of ocean swell, or even panoramas of odors. Cognitive maps, as suggested by the term itself, are *constructed* using various pieces, including the identity and appearance of individual landmarks and their observed relationship to one another. To some extent, it is possible to observe the spatial relationships of landmarks by simply looking at a scene, but such looking must take place from a variety of observation posts, and to generate a metric map—one in which the directions and distances between things are measured accurately—one must know how the observation posts themselves are connected. In other words, in order for us to build a map based on a set of observations, we have to know where we were when we made the observations. The metric accuracy of a cognitive map is going to depend on the accuracy of path integration. Though a gradient map may serve nicely to bring a turtle to a rich bed of ocean vegetation, a cognitive map of the kind envisioned by Tolman and countless others is a much more sophisticated and flexible wayfinding device.

Not everyone agrees with Tolman's conclusions about how his rats solved starburst-maze problems. Many scientists continue to try to understand what kinds of maps can be found inside the minds of rats, but almost everyone agrees that rat minds contain *some* kind of representation of space. Today, more of the debate focuses on questions about *where* in the neural machinery of the rat such maps can be found. In other animals, the debate is more sharply focused on the question of what it might mean to say that an animal's head contains a map. Nowhere has this question been more contended than among those who study the mental world of the honeybee.

BUMBLING BEES

The German scientist Karl von Frisch was awarded the Nobel Prize in 1973 along with Konrad Lorenz and Nikolaas Tinbergen, two others who revolutionized the study of the behavior of animals in their natural habitats. Like Tinbergen, von Frisch attributed much of his interest in animals to a free-wheeling youth spent in the countryside watching things grow and move about. As a scientist, von Frisch conducted clever experiments, much in the manner of Tinbergen's simple studies with digger wasps, that showed how insects can employ landmarks to find their way home.[3]

Von Frisch's most dramatic finding concerned what he called the bees' waggle dance. This peculiar wiggly buzz of movement had first been noted by Aristotle, who speculated that bees used these movements to call attention to themselves.[4] Von Frisch suggested that something much more interesting was happening.

When colonies of bees need food, they send out scouts. When successful scouts return to the hive, bringing nectar with them, other bees leave the hive and head directly for the source of nectar. Von Frisch spent many hours hunched over hives, watching bees return to the nest and then leave again for the fields, and he eventually formed an incredible hypothesis. He believed that the scouts were somehow telling the other bees where the food source was located. Like Aristotle, von Frisch focused on the waggle dance—a stereotyped set of movements in which the scout bee walks in a straight line while buzzing, turns in a loop, walks in a straight line again, and then loops in the opposite direction, making a kind of figure eight. Other bees in the colony follow the scout bee as it carries out this display and, after a few repetitions, observing bees will take flight and head in the direction of the food source. Careful observation of the waggle dance convinced von Frisch that the movements made by the scout bee consisted of a kind of body language that encoded symbolically the location of the food source. He noticed that scout bees that had found food placed at very distant locations carried out longer straight buzzing segments in their runs than those that had found food closer to the hive. So the length of the straight part of the dance signals the distance to the food.

What about direction? The bees that von Frisch studied varied the direction of their dance depending on the bearing of the food from the location of the hive. Von Frisch suggested that the bees were giving a compass bearing to follower bees that was related to the position of the sun. The honeybees that von Frisch studied danced on vertical honeycombs. Von Frisch suggested that scout bees converted direction information such that straight upward was meant to represent the position of the sun. The angle of the dance relative to true vertical, then, was meant to indicate the angle of the route to the food source relative to the sun. There's one more wrinkle to iron out before the bee language that von Frisch described could work properly.

As you can imagine, when these findings were being reported, beginning with von Frisch's studies in the 1920s and continuing to the present day, reactions ranged from skepticism to utter disbelief. Von Frisch was suggesting that an insect with a brain about the size of the head of a pin was capable of acts of communication that could be differentiated from true languages only by the careful hairsplitting arguments of philosophers. Von Frisch himself devoted considerable energy to ruling out simpler explanations for his findings (such as, for instance, that the follower bees were simply following some kind of odor trail to the food source), but only very recently have advances in technology enabled researchers to provide what seems like ironclad evidence for the key role of the waggle dance in bee navigation.

In 1989 a team of researchers at the University of Odense in Denmark built a dancing robotic bee. Although this bee robot did not look very much like the real thing (for instance, it had only one wing and this was constructed from a piece of a razor blade), the team, led by Axel Michelsen, was able to show that not only would follower bees pay attention to the dance of the robot but they would also fly off to look for food at directions and distances described by the robot's dance.[5] As we have seen many times before, the ability to trick an animal into behaving in a certain way by massaging the information to which it is exposed makes for a very convincing demonstration of the importance of this information in the animal's everyday life.

Though it is now well established that bees use the waggle dance to talk about space, controversy still surrounds the question of what kinds of maps of space bee brains might contain. In a widely discussed study,[6] James Gould set out to demonstrate map use in bees using a kind of shortcut experiment. Bees were trained to fly from a hive to a feeder. Once the bees had learned to make the trip from hive to feeder and back again, they were picked up and transported to a new release point. Like Tolman's rats in the starburst maze, Gould reported that many bees flew directly from the release site to the feeder without first returning to the hive. If this route truly represented a novel shortcut, then it suggests that bees possessed a metric cognitive map.

Many researchers took issue with Gould's conclusions. Those who tried to repeat his shortcut findings noted that the bees could complete shortcuts only under conditions where it was impossible to rule out a simpler strategy than map use. Gould's original studies, it was argued, were set up in such a way that certain landmarks were observable both from the hive and from the new release point, so the bees could have been using one of the simple landmark strategies I described in chapter 2.[7]

Until the late 1990s, the tide was turning against the cognitive-map hypothesis for bees, but more recently, the tide has begun to reverse. One breakthrough has been the discovery that bees carry out different kinds of flight missions depending on their history. When young foraging bees are first learning the ropes, they go on what are referred to as orientation flights, whose purpose appears to be to learn the terrain and landmarks surrounding the hive.

Rodolfo Menzel thought that these "orienteering" bees might learn different things about space than experienced foragers. To test his hypothesis, he conducted an experiment in which he first moved an established hive to a new location in order to encourage foraging bees to carry out orientation flights, and then made life even harder for the bees by moving the feeder around from time to time. The feeder was kept close enough to the hive so that it was always easy to find, but because it was never in the same place on two successive foraging trips, the bees would need to pay close attention to their surroundings if they were to find their way home.

Menzel's surprising finding was that bees trained in the more uncertain world of the moving feeder used a flexible navigation strategy based on something like a map. It was as if these bees, on release, took stock of their surroundings, computed their position, and set off on an accurate "beeline" for home. Recent studies using specialized radar methods to track the full flight paths of foraging bees have confirmed this idea. Bees that had undergone only orientation flights were capable of making direct flights from novel release sites to both the hive and to the feeder, and some of these flights took the bees through terrain they had never experienced before.[8]

It is very difficult to conceive how a bee could find its way home from a location that it had never visited before using a completely novel route unless it had a map in its head. Such ability, if it exists, is set apart from all the methods of navigation that we have considered so far. Carrying a map of the terrain, with landmarks and their spatial relationships, within one's nervous system allows one to solve wayfinding challenges at an entirely new level of difficulty, and with the flexibility to solve much more complex and interesting "you are here" problems.

FOOD-CACHING BIRDS

Animals spend much of their time searching for food, securing access to it, and doing what they can to buffer themselves against the vagaries of a perpetually uncertain supply. For many animals this means working carefully to collect and store food during times of abundance so that it can be consumed at a later time, when conditions change. Every young child has been told stories of the prudent squirrel that prepares for winter by gathering and burying nuts to provide a steady food supply for the scarce months ahead, and scientists have demonstrated that such a strategy can indeed be effective in warding off hunger in the bleak winter months.

Some animals store food in "larders" designed specifically to afford defense against theft, whereas others, such as many species of food-storing birds, employ a system of "scatter hoarding" in which they distribute morsels of food across large reaches of the environment for later use. Scatter hoarders have been shown to have a prodigious ability to hide and then retrieve food, and so have been subject to intense scrutiny by scientists interested in how animals use space.

Chickadees, tits, and nutcrackers are among the species of birds that store food in cache sites. Some of these birds have been observed to use up to 80,000 different cache locations in a single fall season. Seeds, decapitated insects, or bits of worms are stored in a wide variety of locations, such as under tufts of dirt on the ground, underneath the bark of trees, and inside hollow plant stems. Cache sites can be either right on the ground or high above it, and birds can cache food either very close to where it was found or at distances of up to about 100 meters away. Most cached food is recovered and consumed within a few days of storage, but some morsels are known to have been hidden away for some months before being retrieved.[9]

How do we know that the birds actually remember all these cache locations? It would be easy to imagine that they might use some kind of simple rule to find cache sites. For example, if a particular bird always stores food in hollow stems, then it might find its own caches by searching randomly among all the hollow stems that it encounters. I use a similar strategy when I discover, and recache, Halloween candy in my house. Knowing that almost all my children are more vertically challenged than I am, I cache tasty treats on high shelves. But given the sad state of my overtaxed memory, I seldom remember *which* shelf holds my stash. Luckily for me, it doesn't take too long to search all the high shelves in my house for a sweet payoff.

We know, however, that food-storing birds don't use this kind of rule-based strategy because of some experiments in which field researchers watched carefully where birds stored seeds and then went out and stored extra seeds in nearby locations that looked almost the same as the original cache site. When the birds revisited cache sites, they showed a strong preference for their own sites as opposed to those prepared by the researchers. All evidence suggests that these birds actually remembered the locations of their cache sites in great detail, days or possibly even months after they stored the food.

Such an extraordinary memory for spatial locations presents myriad opportunities for us to understand how warm-blooded animals manage space. As some birds can also be coaxed to cache and recover food in the artificial setting of the laboratory, scientists have been able to learn a great deal about how spatial memories in food-caching birds are put together.

In one experiment, Clark's nutcrackers were trained to dig in a specific location on a sawdust-covered floor in order to unearth a small stash of pine seeds. The location of the stash was reliably indicated by a set of prominent landmarks, but the setting was cleverly arranged so that the birds would need to be able to consult a mental map relating the landmarks they could see from the target location to a set of screened-off landmarks that they had learned to associate with the target during training but could not see when near the target site. The nutcrackers were able to accomplish this difficult task, suggesting the existence of a cognitive map, but their accuracy was lower than expected, leaving some lingering doubts about whether the birds possessed a cognitive map or whether they were falling back on some kind of simplifying trick to find the target.[10] But even if future studies show that the bird map is not quite up to the lofty standards set by Edward Tolman and others for a cognitive map, birds have shown great flexibility in solving some difficult navigational problems in the face of virtually every curve that wily experimenters have been able to throw at them. These animals cling to their place on the planet with remarkable tenacity, relying on one backup system after another in order to avoid becoming lost.

It is difficult to imagine that anyone hearing these accounts of scurrying rats, soaring bees, or industrious food-caching birds could harbor any lingering doubts about whether such feats of wayfinding prove the existence of mental maps, yet in some quarters debate still rages. To find one's way from any location to any other, it is necessary to compute both distances and angles. In other words, the bee, bird, and rat maps that I have described to you work properly only if they are based on the same kind of geometry that we use to design street maps of cities. Not everyone agrees that animals such as birds, bees, and rats possess such maps. Indeed, many argue that in trying to find convincing evidence for such maps in the minds of animals, we are simply asking the wrong questions. What would be more fruitful, they say, would be to understand what sense animals need to make of space to maintain their habits of life, and then to do the experiments necessary to see how such spatial sense is constructed.

Strangely, fewer of us find reason to doubt that there are maps in the heads of human beings. We will see in the next chapter that such maps are peculiar kinds of things that cut close to the heart of the unique human connection with physical space.

CHAPTER 6
MUDDLED MAPS IN HUMAN MINDS

THE PECULIAR NATURE OF OUR MENTAL MAPS AND
WHAT IT SAYS ABOUT HOW WE UNDERSTAND SPACE

I see nothing in space as promising
as the view from a Ferris wheel.

E.B.WHITE

On one level, it seems strange to doubt the hard realities of
space and time. If I want to know how long it will take me to
drive to Chicago, or whether there's still time to walk to work, the
relevant calculations are straightforward and are based on equa-
tions that have been well understood since ancient times. The same
basic mathematics transported tiny capsules full of men across the
vast reaches of space to the moon. As he stepped onto the dusty
surface, nobody heard Neil Armstrong ruminate on the possibility
that the massive Saturn V boosters that had pushed him and his
shipmates across the void had been a figment of their minds. The
beautiful blue planet they looked back on from their lonely vantage

point in space was no invention. It was real and it was a long way off. Doubting any of this seems a regressive step.

Yet think of the everyday phenomena of your own lived space. If space is nothing more than an infinite expanse of pure geometric nothingness, then how can the distance *from* home when departing so often feel longer than the distance *to* home when returning? Walkers gauge the distance between two points based on the number of turns they have made along the way rather than the distance they have walked or the time spent walking. When placing new furniture in our home, why are we so often surprised by what does and doesn't fit into spaces that can be almost as familiar to us as parts of our own body? In earlier times, we might have attributed such strangeness to the mystical power of place. Now that we see places more as objective locations studding cold space than as powerful entities brimming with their own special kind of life, we turn inward for answers to such questions. We seek answers in our own psychologies.

Given the oddities of the psychology of space that I've just listed, along with many others that I'm sure you can think of from your own life, we should be prepared for the possibility that whatever mental maps we might possess may have features not predicted by the impressive wayfinding feats of the birds and the bees. Not only can we expect our cognitive maps to be inaccurate but at times they may turn the world into a place that defies the laws of physics (or geography at least) in ways that are downright weird. In spite of this weirdness, such maps often seem to work. The sketch maps that we draw to help guide visitors to our house may have the faintest of resemblances to real geography, but what is most important about such maps, mental or otherwise, is that they serve the purpose for which they are designed. Provided that they work, there may even be some advantages to constructing maps that make only weak connections with physical reality. They may be easier for us to remember, or they may leave out details that might confuse us. It may even be that our ability to play fast and loose mental games with space is what underlies many of our most dazzling cognitive feats, especially those that free us from the bounds of real spaces to allow us to inhabit those made of electrons. The maps that we imagine and draw may provide useful signposts to the organization of our own spatial mind. When we sketch a map, what we include and what we leave out may have much to teach us about the cartography of our own inner mental spaces.

THE ORIGINS OF HUMAN MAPS

Archaeological evidence suggests that the production of primitive maps predates written text and numbers by several thousand years. In a way, this is not surprising, as there seems to be something deep and universal about our desire to find ways to represent spaces to each other. Though children need some instruction in order to understand how to use a map, it is relatively easy for them to grasp the basic idea that a drawing is related to a physical space in an orderly way. Such a cognitive leap is much simpler than, for example, coming to terms with the idea that text—long lines of abstract squiggles on a page—may convey meaning. The cognitive building blocks that children need to appreciate the connection between topological aspects of real space (what is connected to what) and map topology emerges at an early age, certainly in the preschool years. An ability to use maps to solve metric problems, such as deciding on the shortest route from one location to another, arises much later in life, if at all. Much evidence suggests that for most of us the ability to represent space with metric accuracy may never develop fully.[1]

The earliest known maps, scratched into stone about 40,000 years ago, were simple depictions of natural objects, having more in common with prehistoric rock art than with topographic maps. Though some of these depictions look as though they may have been maps, we know little about their function, so it can be difficult to be certain that they are maps at all. For something to be a map as conventionally understood, it must have an intended use that relates to either understanding or moving through geographic space. For example, if I drew a picture of my neighbor's face, it would be true to say that the picture was a "map" of her face in the mathematical sense, but that is hardly what we are thinking of when we use the word *map*.

Some early carvings found in parts of Europe and North Africa portray scenes of animals and a few stylized figures that may or may not represent parts of the spatial context in which the animals were found. Later carvings are much less ambiguous and show two of the common hallmarks of modern maps. One is a system in which repeated use of a symbol is made to stand for some feature of an environment. The symbols are thought to represent dwellings, individual people, and sometimes animals. The other cartographic hallmark of these very early drawings is that some of them show survey, or overhead, views of the environment. In some cases, perspectives are mixed up so that some figures are shown in profile while others are shown from overhead.[2]

Children's drawings also show such mixtures of perspectives, as do certain styles of modern art—cubism for one. Children may mix perspectives because they are unable to put together accurate maps of spaces from viewpoints that they cannot adopt themselves, whereas visual artists using cubist techniques are motivated by the desire to show objects from unusual, impossible, or multiple perspectives. We have no way of knowing what prompted our ancient ancestors to use such multiple perspectives in their drawings. Is it possible that in such mixed perspectives we are witnessing an early struggle to break free of what can be seen directly in favor of the spaces of imagination?

There is much that we can never know about the drawing skills, the cognitive toolbox, or the motivations of early primitive artists, but we can certainly see the beginnings of the struggle to comprehend and capture on stone the properties of large-scale geographic spaces. Survey drawings of prehistoric environments suggest that some early humans were capable of adopting a perspective that may have been impossible for them to have seen with their own eyes.

Much later, there can be little doubt that people produced drawings that were meant to be appreciated from a bird's-eye view. One of the most famous examples of such representations was made by the Nazca of Peru. The famous Nazca lines, carved out of the ground more than 2,000 years ago, portray a variety of objects and animals on such a grand scale that they can be viewed in their entirety only from a great height above the ground. Although it seems likely that these drawings were meant to be observed by deities, a number of other theories have been entertained, such as that the drawings were giant calendars, irrigation channels, even landing pads for extraterrestrial spacecraft![3] These depictions were not maps, but they illustrate an increasing facility with the geometry of geographic space.

One thing that makes it so hard to know whether some of these early graphics were real maps is that we rarely have any idea how the maps related to the physical landscape as it appeared at the time the maps were drawn. The spans of time involved are so long that much would have changed between the time the lines and marks were hammered into stone and the time they were unearthed by modern scientists. And even if these drawings were meant to represent the physical landscape, it isn't clear that they would have been used as aids for navigation. Some may have had symbolic or religious functions in the lives of those who created them. The detailed maps of what seem to be agricultural fields drawn by some cultures may have served as a kind of amulet designed to fulfill a superstitious belief that representing the plots of crops would help to ensure a successful harvest.

Though we have to be guarded in our interpretations of their meanings and uses, ancient maps provide an important window into the origins of human representations of space. Regardless of the many questions about these maps that we will never be able to answer fully, these artifacts leave little doubt that an important hallmark of the unique human engagement with space consists of this basic tendency for us to construct mental models of the stuff. Though these models can have all kinds of uses, it is only in rare cases that they have anything like metric accuracy. Much modern research has been devoted to exploring why this might be.

GEOGRAPHY LESSONS

Time for a pop quiz:

1. Which city is farther north: Seattle or Montreal?
2. Which city is farther west: Los Angeles or Reno?
3. Imagine the map of North and South America. Which North American city lines up with the west coast of South America: Vancouver or Chicago?

Before I give you the answers, think about how you imagined geographic space. The questions deal with areas of land that are too large to have been visualized all in one go (unless you happen to be an astronaut), so your images would have been based on maps that you have seen rather than from direct personal experience. In a way, these questions might seem to have less to do with the way we find our way through space using mental maps and more to do with how we imagine the space and geometry contained in simple pictures. Perhaps it doesn't even matter that the images happen to be maps at all. As you'll see, though, there are some interesting connections between the ways that we imagine the kinds of pictures conjured by memories of maps we've seen many times since childhood and the ways that we try to use our own cognitive maps to navigate through space.

And now the answers, which may hold one or two surprises for you:

1. Seattle is farther north than Montreal.
2. Reno is farther west than Los Angeles.
3. Chicago, and not Vancouver, lines up with the west coast of South America.

If you're like many of the participants in studies of how we form cognitive maps, then at least one of these answers is likely to send you off to find the nearest atlas to check the facts, but I can assure you that you will find no errors.

Barbara Tversky, a cognitive psychologist, carried out groundbreaking studies on cognitive mapping at Stanford University in the 1980s.[4] These studies showed some of the key differences between mental maps and the physical spaces they represent. One such difference arises from what Tversky calls alignment. The idea is a simple one. Our mind's eye tends to shy away from the diagonal, the oblique, the slanted, in favor of horizontals and verticals. Hence, when we represent a line or surface with such irregularities, we tend to straighten it out. This accounts for our tendency to align the edges of North and South America when we imagine them on a map, so that we are surprised when we are told that the western edge of South America aligns with a city in the eastern half of North America.

It isn't just our impression of large-scale geographic space that is influenced by alignment, but also the smaller spaces that we traverse every day. Erik Jonsson, a retired engineer with a lifelong interest in navigation, conducted an informal study of drivers stopping at a rest station on an interstate highway in the United States. Though most drivers knew where they had come from and where they were going, few of them could tell Jonsson their compass bearing just before they had pulled off the highway. They had straightened out all the curves in the road.[5]

Tversky conducted another study in which she asked Stanford students questions about locations close to the campus. Just as we rotate whole continents in our heads to make them line up, the students tended to align local bodies of water with north–south and east–west compass headings. This tendency was so pronounced that the hapless students ended up making considerable errors in judging the relative positions of local landmarks and communities that they visited every day.

What causes us to rearrange our maps of space to conform to tidy horizontal and vertical lines that exist only in our imagination? Tversky argues that one part of the answer has to do with the natural axes of symmetry of our bodies, and the fact that we are upright beings who spend much of our waking lives orienting ourselves to the force of gravity. We tend to line up the world with the horizontal and vertical because two of the most salient benchmarks against which to compare all spatial attributes are the force of gravity and the appearance of the horizon. It is certainly true that preference for the vertical and horizontal is written into everything from our mythologies of space and the design of our street maps to the basic operation of our visual system.

Noted geographer Yi-fu Tuan describes a deeply rooted tendency for human beings to categorize the cardinal directions of space according to the axes of the body.[6] Most cultures make clear distinctions between the meanings of "in front" and "behind" as they refer to the orientation of the body. The various cultural connotations of "left" and "right" are reasonably well known. In most cultures, the left side is inferior to the right. We shake with our right hands; we eat with our right hand in cultures where utensils are eschewed. Our most valued assistant is our right-hand man. The left hand is still referred to in the scientific literature as the sinistral hand, in contrast to the right, dextral hand, with the modern word *sinister* taking its root from an early designation of left-handedness. One interesting exception to the low status of the left is found in Asian cultures, in which the primary cardinal direction is the south (the main entrance to the Forbidden City is on its south side, for instance). This being the case, left is the side of the rising sun, and so is afforded privileged status.

In the ancient Asian science of feng shui, the orientation of buildings and cities with respect to the cardinal directions was considered to be critical to the health and success of the built environment. Streets and buildings were aligned carefully with such directions, and the human body can be seen in microcosm in some applications of feng shui principles. As our straight, two-legged posture pushes our heads up against the force of gravity, it is a cultural universal that physical height correlates with the direction from which human power emanates. Palaces tower over landscapes for reasons that transcend military strategy. The city of Beijing is built on largely flat terrain, yet the Forbidden City, the ancient throne and power center of the old city, is considered to be at the peak, with the surrounding regions arrayed below it like a set of virtual terraces.

Architecture's embodiment of power structures, equating height with might, did not end with the decline of palaces, royalty, and feudal systems of governance. It takes only a quick glimpse at the skylines of the Manhattans of the world to be convinced of this. Gigantic, geometrically perfect towers of trade and commerce jut above the horizon, fierce and warlike. It is beyond coincidence that the modernist World Trade Center was the target of a vicious terrorist attack in 2001. Mohammed Atta, the apparent mastermind of the murderous assault, was trained as an architect and worked as an urban planner. One report suggests that he was driven into the arms of the jihad in reaction to the architectural westernization of parts of Cairo. Atta, more than many others, would have understood and resented the symbolic significance of the towers.

At the much more mundane level of the everyday operation of our sense of vision, it has long been known that we are more sensitive to vertical and horizontal lines than to oblique ones. Sensitive vision tests using fine grids of lines show that we are better able to see spaces between lines when they are oriented either vertically or horizontally than when they are presented at oblique angles. This sensitivity bias for aligned stimuli has to do with a preponderance of neurons in our visual brain that are tuned to the horizontal and the vertical. Nobody knows the origin of this bias, but it is likely to be related to the orientation of our bodies with respect to gravity.

Not only are we more sensitive to horizontal and vertical lines but we also seem to prefer them in images. Visual artists such as Piet Mondrian certainly understood such preferences. His paintings, mostly consisting of grids of colored blocks, were devoid of the oblique. Mondrian was a member of an artistic group known as the de Stijl group that tried to place aesthetics on a scientific basis. An explicit part of the artistic manifesto of the de Stijl group was the dictum that oblique lines must be avoided at all costs. Modern psychological experiments have confirmed the intuitions of these artists. When participants in an experiment were asked to judge the pleasantness of a series of Mondrian paintings presented either as the artist had intended or tilted to oblique angles, they preferred Mondrian's horizontals and verticals.

Our preference for vertical alignment crops up in everyday life as well. Who among us has not felt (and perhaps succumbed to) the desire to straighten a crooked picture on a stranger's wall? Filmmakers use a technique, pioneered by Alfred Hitchcock, in which slight camera tilts produce sensations of visceral discomfort in viewers. When lines that would normally follow the horizontal or vertical are pushed slightly toward the oblique, cinemagoers are, quite literally, put on edge.

Note what is happening here. Psychological tendencies that have their roots in the orientation of our body and the organization of our senses seem to have taken precedence over what we see and feel of the dimensions of physical space. When we try to imagine such spaces, whether by remembering pictures or by mentally replaying our movements through space, our predilection for the clean vertical and horizontal contours overrides much of our ability to represent spaces accurately. We construct spaces rather than sense them.

CHUNKS OF SPACE

Anyone who has looked at children's drawings knows that our mind tends to simplify visual patterns. The complexity of real forms comes to be replaced by simple collections of basic shapes, organized according to schemas—sets of rules that dictate how these basic shapes must be fitted together. Such schematization of shapes results in human figures that become sticks with big round heads. Birds in flight become stylized squiggles, and the sun is represented as a yellow disk with a symmetric burst of sunbeams around it. An important part of the training of an artist consists of learning to draw what is *seen* rather than what is in one's mind. In other words, artists must be taught to unlearn the routines of schematization in order to draw accurate copies of real life.

The same processes that simplify our drawings act on our conceptions of space, and the reasons are similar. Maps that simplify spaces, straighten curves, even out distances, and turn irregular clumps of land into orderly geometric shapes are much easier for us to remember than those that contain every warp and wobble of raw geographic space.

One of the best examples of this tendency to regularize space, and another of the factors responsible for the geographic illusions that I pointed out in my list of questions, is something called regionalization. To understand how regionalization works, try this exercise: as you are sitting at this moment, close your eyes and point to the location of some object that you know to be in the room with you—a lamp, window, or door, it doesn't really matter. If you open your eyes, you'll probably discover that you were able to point reasonably accurately to the object you chose.

Now try something different. Imagine an object again, but this time, choose something that is outside the room you're sitting in. If you're sitting in your home, you might try pointing to the position of an object in a room on another floor. Now try pointing to the location of your best friend's home. What about the location of city hall, or the nearest body of water? As you go through this exercise, you'll notice that some targets are easier to point to than others, and that completing this task for certain kinds of landmarks will involve considerable mental gymnastics. You may have noticed something else as well. When we are required to imagine the position of an object that is not currently within view, we try to reconstruct that object's position in a series of discrete steps. I can point to the lamp across the room with no difficulty whatever, but if I then try to point to the location of the toaster in my kitchen, several rooms away, I try to reconstruct a path from where I am to the goal object, and then mentally add together all the segments of the path. Instead of trying to conjure a spatial image of the toaster from my current position, I manage the problem by first imagining the view from the threshold to my study as it opens into the hallway beyond. This orients me toward the kitchen. Then I imagine myself at the end of the hallway, looking into the kitchen. At each stopping point I imagine a view, and then I try to add all the views together to connect the beginning to the end of my path. As you can imagine, the larger the number of segments involved, the more likely we are to become inaccurate.

In one study of the psychological regionalization of space, people were taken into a small, windowless room and, after being given a chance to look around to become familiar with their surroundings, they were asked to close their eyes and point to objects in the room, much as I suggested you do a few moments ago. Participants found this task easy and generally did well at it. Next, they were taken out of the room and led on a walk around a part of the building that surrounded the experimental room. At various stopping points on their walk they were asked to point to objects inside the room they had left. Once back inside the room, they were asked to point to objects they had seen outside. Participants found both of these latter tasks to be much more difficult, presumably because they were trying to piece together views of space in the same way I did when trying to imagine my toaster. Other studies conducted with people in spaces that they had used regularly for at least two years (their offices) showed similar findings, so our difficulties in making spatial connections do not seem to have much to do with the amount of experience that we have with a part of space.[7]

Why do we regionalize space? One part of the answer is that regionalization is a handy way to help out our limited memory resources. Most people, when called upon to remember a long list of items, will resort to a strategy called chunking. For example, when I try to remember a grocery list, I categorize items into different groups—fruits, vegetables, meats, dairy, and so on. Then my job is to remember a small group of lists, each containing a handful of items, and this is easier than trying to remember one long list. The process of learning about spaces is similar. It is easier to memorize the locations of a series of objects within each room, and also to remember the rough layout of a set of rooms in a house, than it is to try to remember a long set of locations of objects and to place them all on one very large map.

Although this kind of hierarchical organization of space can help us to manage our memory load, it results in distortions in our maps of space. In our mental maps, the distances between points that are in different regions seem longer than distances between points within the same region. This effect is strong enough that it not only affects how we think about our lived spaces while sitting idly and reflecting but it also influences our choice of routes when walking or driving. A route that requires many changes in direction seems longer to us than one that is straight. The reason is that each turn brings into view a new set of features, and so constitutes a new region. The practical implications of these effects have drawn the interest of those who design buildings, neighborhoods, and cities because it is possible to influence how people might use a space by judiciously tinkering with its size and shape. In an urban center, if we want people to get out of their cars and walk, many clever tricks of planning will entice them to do so, but one tactic is to make interesting locations *appear* closer together by prudent organization of space.

Our tendency to chunk space into regions is a cornerstone of our spatial mind. Even experiments that use highly abstracted spaces show a strong effect. For example, one study showed that if research participants were asked to memorize the positions of a random array of objects on a computer screen, they did so by mentally dividing the screen into a series of regions based on the locations of the objects they saw. When they were quizzed about the distances between objects, distances *within* mental regions shrank compared with those that *crossed* regions.[8] Most of us, when asked whether Seattle or Montreal is farther north, will answer by remembering that Canada is north of the United States and then assume (erroneously) that Montreal is north of Seattle. Similarly, Reno seems as though it should be east of Los Angeles because Nevada is (partly) east of California.

Mental maps, like the maps we sketch on napkins to guide our friends around town, are filled with inaccuracies, distortions, and even absurdly impossible spaces. When we begin to draw a map and we preface our effort with words such as "This is not to scale, but … ," we are implicitly acknowledging this feature of the weird spaces that we share with one another. But beyond being a kind of graphical shorthand we use to convey the main features of geographic spaces, the maps we sketch have deep affinities with the properties of the spaces that inhabit our minds. Our minds treat distance and direction with cavalier disrespect but represent topological relationships with greater clarity. Though we don't seem to have much of a grasp of how far away things are, nor what their angular relationships might look like (especially when they cross regions), we have a good idea of how different parts of space (roads, paths, hallways) are connected. Just as topological maps, like twisted rubber sheets, can tolerate much distortion while retaining some information about spatial relationships, so can the maps we hold in our heads help us through spaces, particularly the ones that we build for ourselves. Provided we understand how regions are connected, and what is in each one of them, we can plan routes to goals. We may not always (or even often) take the most efficient route, but we usually know which set of spatial decisions will help us get to our final destination. I might not know how far it is from the bakery to the post office, or which one of those goals is farther from my house, but I know how to get between one and the other, and I know how things will look when I get there.

Are these topological maps of space the kinds of cognitive maps that Edward Tolman had in mind in the 1940s? Are they the kinds of maps that researchers in animal behavior have fought over for the intervening sixty years? Not really. Rats solving a sunburst maze, bees interpreting a waggle dance, and pigeons solving a gradient map to find their way back to a roost need access to a map that retains some metric features. Novel shortcuts work only if we understand the real distances and angles between locations. All too often, unless we are trackers using traditional methods, or members of a culture living by our wits in barren landscapes where one wrong turn can kill us, such spatial information eludes us. Unlike other animals, which are tightly anchored, body to ground, fixed to the earth with a sureness of footing that can be almost impossible to sunder, human beings seem preternaturally prone to a kind of spatial flight of fancy in which our minds sculpt physical space to suit our needs. Though under certain circumstances and with specialized training we are capable of some prodigious feats of navigation, the more common occurrence for modern human beings is that we flounder through a highly schematized version of physical space that has only a weak relationship with the real thing. When this strategy works for us, it is often because we have designed an environment for ourselves that is replete with spatial crutches, an environment that makes heavy obeisance to the metric inadequacies of our spatial brain. But when the strategy fails, it can do so quickly and disastrously, sometimes even costing us our life. We spend much of our life being only one miscue away from complete spatial disorientation.

SPACE AS MENTAL PROJECTION

The news is not all bad for human beings. Though our mind is put together in such a way as to make accurate maps of larger-scale spaces a difficult conundrum for us, the same cognitive capacities that make us get lost walking to the corner store may underpin some of the most remarkable features of our mind, including those that set us apart from all other animals. Our ability to abstract ourselves from our current spatial context, to close our eyes and to visualize ourselves in some other space, no matter how stylized it might be, is probably a uniquely human thing. Our ability to visualize the floor plan of a building from an overhead perspective and to not only see ourselves in the building but to see what we would see from our *imagined* position is a capacity possessed by no other animal. Though astral projection and other out-of-body experiences may be the stuff of fantasy and science fiction, our ability to throw around our viewpoint at will is real and significant.

Jean Piaget, one of the founders of modern developmental psychology, devised a task that he called the three mountain problem. Children were shown a model of a miniature landscape containing what looked like three mountain peaks. They were asked questions about the appearance of the peaks from alternative points of view. For example, a child might be asked to describe what another child, sitting facing the first child, would be able to see from his position. Piaget found that before a certain stage of development, children had great difficulties with this task.[9] In a very real sense they were locked into their own point of view. Older children can adopt these alternative views much as adults can. For me, the wonder of the three mountain problem is not that young children *cannot* do it but that adult human beings *can* do it. This means nothing less than that, for all our topological weirdness, we have found a way to free ourselves from the confines of physical space and to take flight. While my body is planted here in this chair before my computer screen, my mind can be down the hall in the kitchen, down the road on the beach, or looking down from high in the sky. From each of these vantage points, I can estimate roughly where to place my own faraway body in the picture.

No less revolutionary than our ability to picture ourselves in physical space from alternative points of view is the fact that we can picture a place that does *not* contain us. Even though I am no longer in the kitchen, as I was a few minutes ago (stealing one of the few remaining home-baked chocolate chip cookies before the children came home from school), I am completely confident that the kitchen still exists. Even though I have not been to the beach now for two days (sigh), I know that it is still there and that it doesn't somehow slip off the edge of the universe when it is out of my graspable space. Though these facts might seem obvious, their implications are profound. Were it not for our ability to apprehend parts of the physical world beyond our immediate sight or grasp, we would be very different kinds of beings. An essential ingredient of self-consciousness (and here I mean not the awkward, gangly self-consciousness of a shy teenager but the objective awareness of oneself as a causal agent in the world—in other words, as a being who can make things happen) is this ability to abstract space. If we could not take a vantage point on the world that was outside our own body, we could not appreciate that the physical world endured when it was outside the range of our senses, nor could we appreciate the difference that it makes that our own body is in that world. Understanding the difference between a world that contains us and one that doesn't is at the very heart of what it means to have a personal identity. Without the ability to take an objective perspective on space that is not centered on ourselves (understanding that the world goes on without us, in other words), it is hard to imagine that we could have much notion of the passage of time, either. Psychologically, time is inextricably bound up with movement. The horizon represents a place of the future or, if we turn around to see where we've been, the distant past. Without an idea of time, it is difficult to imagine being able to see oneself as an enduring being with a personal history. The self that binds all of its history into one cohesive biography can do so only by using time as the glue.

———————

Maps can either be based on some kind of physical gradient in the world or they can be constructed mentally by careful observations of landmarks and measurements of the distances between them. Pigeons, food-caching birds, bees, and many other kinds of animals appear to navigate with stupendous accuracy by relying on one or the other of these two types of maps.

In this chapter, we have found that human map construction and use suggests that our own mental spaces are composed of a strange, rubbery substance. Though most of us can find our way home every night, we often have little cartographic insight into how we got there. We live teetering on the brink of spatial collapse, but we're made blissfully unaware of this by a plethora of wayfinding aids offered up by architecture and modern technology. The maps of space that reside in our mind, though they are nothing like the spaces described by physicists or mathematicians, represent a kind of compromise between our need to conquer space well enough to survive and the limited capacities of our memory. What we cannot perceive directly or remember, we invent. The silver lining of this act of invention is that our ability to imagine, stylize, and transform space with our mind frees us from it in a way that is unique to us. This freeing of our mind from the trappings of physical space has been one of the key ingredients in an evolutionary path that has helped make us into beings, unique among all living things on the planet (and perhaps in the universe) because we can both imagine ourselves being elsewhere and imagine an infinity of "elsewheres" existing without us. The same regionalization that mentally disconnects us from other spaces allows us to free ourselves from the constraints of physical space in a way that is impossible for any other animal. By inventing space, we have made it our own.

Our minds hold a strange and wonderful power over space. Unlike birds, bees, and other creatures of field and forest, we seem able to make spaces, bend them to our needs, and imagine them as things other than what geometry suggests they are. It may well be true that our ability to conjure space in this fashion has been one of the major engines that has helped to push humanity to a pre-eminent position as the only truly self-conscious being on the planet. The combination of a mind predisposed to take in complex views using a highly developed sense of vision and then to join those views together in an odd amalgam based on topological connectedness has allowed us to invent and construct spaces beyond the wildest imaginings of early human beings.

But it isn't just the construction of our minds that has enabled us to slip off the mantle of the geometry of space. Our soaring ability to harness energy and technology has allowed us to construct our physical environment in almost any way we please. Hand in glove with our spatial mind, we have used our abilities as toolmakers to design environments that support and extend our mental penchant to transcend physical space. Everything from architectural design through urban planning to modern light-speed communication technologies has been designed to reflect, support, and extend our mastery of physical space.

PART II

MAKING YOUR WAY IN THE WORLD TODAY

HOW OUR MIND SHAPES THE PLACES
WHERE WE WORK, LIVE, AND PLAY

CHAPTER 7
HOUSE SPACE

HOW OUR MENTAL MAPS INFLUENCE OUR
BEHAVIOR INSIDE OUR HOMES

When the peaks of our sky come together,
my house will have a roof.

PAUL ELUARD

An old mentor of mine, something of a father figure, once told me that the two numbers that have the most impact on our economic future are the starting salary of our first job and the price that we pay for our first house. At the time, I had no reason to believe that I would ever have a job or a house, so I just nodded politely without paying much attention. When I finally bought a house, I had little idea what it was really worth and, within days of moving into it, less of an idea what had made me buy this particular house. A small river ran through the basement. The floors were so uneven that a baseball placed on the kitchen floor would roll quickly to the other side of the room. The "yard" consisted of

about a half acre of prime wetland—not much good for gardening and requiring some vigilance on my part to ensure that neither children nor dog were sucked into the bog. What I did know
about that old house, though, was that I loved it from the moment
I set foot in it, even as the hungry fleas from the previous owner's
dog leapt out of dormancy in the old carpets to attack my ankles
for a fresh meal.

Years later, when I was selling that house, I still understood
very little about what made people buy houses. I placed all my trust
in the hands of a realtor, who handed me tip sheets filled with nug-
gets of wisdom for prospective sellers. Most potential buyers, I was
told, have made a decision within eight seconds of getting to the
front door. Hence, I was advised to polish the doorknob to a high
gloss and to make the entranceway immaculate. At open houses,
one of the most commonly inspected areas is the inside of the air
exhaust fan over the stove, I read. I had no idea why one would stick
one's head into such a nasty little space, but I dutifully scoured my
exhaust fan for all I was worth. Not only this, but when looking for
a house to buy, I developed a new habit of peering up under stoves,
always wondering what surprises might be in store for me.

The new trend in the house-selling business is what is called
fluffing or staging. Teams of experts, self-described design gurus and
fashion divas, descend upon one's house, sometimes removing all
the contents and replacing them with what amounts to a Hollywood
stage set. Though it is difficult to get one's hands on meaningful statistics with enough detail to allow a scientific opinion, the consensus
among real estate professionals is that such staging reaps enough
additional profit for the seller that it would be worthwhile at twice
its cost. (Though the price of a good fluff can be modest if one is
willing to do much of the labor oneself, a full going-over on a large
estate can easily cost tens of thousands of dollars.)[1]

Much of the fluffing effort is devoted to obvious cosmetic concerns such as decluttering, euthanasia for the garage-sale armchairs that lurk in family rooms, and sometimes a fresh coat of paint in a neutral tone. But some of the staging effort consists of a deliberate attempt to sculpt the perceptions of the potential buyer with respect to the configuration of the interior spaces of the house. By artfully placing a few good pieces of furniture, not only can one exert a strong impression on the viewer's assessment of the size and shape of a space but one can manipulate the manner in which a space is explored and scrutinized.

We seldom pay as much attention to how we perceive and use the spaces in a house as when we are preparing it for sale or when we are considering a purchase, yet the configuration of built spaces in our residences, offices, and the institutional buildings that we frequent always exerts a powerful influence on our behavior, how we move about, where we linger, and how we feel. Good designers understand some of these influences intuitively. The scientific facts of spatial cognition, the manner in which our brain is put together to perceive and act in space, and the spatial peccadilloes of the human mind can help to shed considerable light on these processes.

ON KNOWING ONE'S PLACE

Some time ago, when I received a pleasant bundle of cash (in fact it was part of the advance that I was paid for writing this book), I treated myself to something that I had always wanted—an exquisitely comfortable leather reading chair. When the chair was delivered, I was immediately confronted with a difficult decision: where to put it? Some locations were ruled out on the basis of inadequate light or inconvenience. Others seemed to jut impractically into areas that I knew would be frequented by noisy, playing children. One location, buried in a small room, nestled against a sheltering wall, seemed perfect. I placed the chair in the appointed spot, set up a small side table to hold a book and a glass of wine, and purchased a stylish reading lamp to place at my elbow. For the next six months, I used the chair three times, instead preferring to do all my reading in another, less comfortable chair situated in the middle of what was not only the largest room in our house but also the one that was most likely to be filled with children, toys, and, because this room was open to the kitchen, clanging pots and dishes. I've now moved my new chair to another location, near a window, facing into the big room, where it gets a bit more use but still not as much use as the old armchair in the middle of the large, open space.

What influences where we rest in our homes, where we place the furniture, and how we spend our time? To some extent, we are steered by simple pragmatics. Bathrooms must be used for taking baths and showers, kitchens must normally be used for preparing meals, and beds must, by definition, be placed in bedrooms. But much of our quiet, waking time is flexible and influenced by simple preference. How are such preferences formed? Can we make scientific predictions about where people might be drawn to walk and linger inside their homes?

Figures 6 and 7 show a part of the floor plan of my house, with the locations of the two chairs—the expensive but seldom-used new chair in the reading room and the heavily used old chair in the family room/kitchen. Theoretically, the smaller room is supposed to be a quiet room for reading, but in practice it often seems to fill with Lego blocks and children's shoes. I don't know why. Beside the smaller room, and opening into the family room, is a small hallway that leads to the side door.

Consider the walls of the rooms to be nothing more than barriers to visibility. These barriers set up two different kinds of contours. Closed contours are just the physical walls, usually the exterior walls, beyond which we can neither move (without going out a door) or see (except through windows). Regardless of where I might be sitting in the space, the positions of the closed contours do not change. Open contours, on the other hand, are those that are set up by the positions of interior walls. So, for example, the partial walls that separate the family room from the reading room set up barriers to visibility that vary depending on where I sit or stand. For example, as I move about in the family room, the amount of space inside the reading room that is visible to me will change.

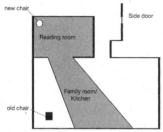
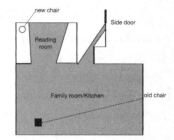

Figure 6: Field of visibility from new chair   **Figure 7:** Field of visibility from old chair

In each figure, I have shaded in gray all the parts of the house that I'm able to see from the vantage points of each chair. These regions of visibility from a particular vantage point are referred to as isovists.[2] As you can see from the examples, the isovist is formed by both the closed contours of the outer walls of the house and the open contours caused by limitations on visibility (the simple fact that, unlike Superman, I'm not able to see through walls). The very simple isovist analysis of my armchairs makes it easy to see a couple of salient features of my favorite seating position in my house. First, the location of the family-room armchair offers the largest available isovist in my entire house. I can see all of the family room and kitchen, a good part of the reading room, and some of the hallway. What I haven't drawn in the diagrams is a row of large windows showing views of the back yard, a large part of which I can also see from the family-room armchair. In contrast, the isovist that is available from the reading-room armchair consists only of the reading room itself and a thin wedge of space in the family room. Without doing any very sophisticated analyses of the size and shape of the isovist, it is pretty obvious that one reason for my preference for one chair over the other is probably the size of the isovist it offers. Another is that the location of the preferred chair is virtually the only position on the main floor of the house (other than in the hallway itself) from which it is easily possible to monitor the comings and goings of the side door, the most commonly used door in the house. From the standpoint of security, this is another reason why it makes sense for me to prefer to sit in the old chair rather than in the more comfortable chair in the reading room.

This simple analysis of a part of the interior space in my house has revealed some of the reasons for my own patterns of lingering and movement. It has long been known that one of the features that attracts us most strongly to interior spaces is a feeling of spaciousness, so it makes some sense that we gravitate to the largest isovists we can find. Experiments have shown that we are preternaturally skilled at finding these large isovists. When participants in research studies were put in architectural spaces and then asked to find the center of the environment or the position from which most could be seen (or, conversely, the best hiding place), they moved quickly and accurately to the correct location. We can parse space very effectively using views, vistas, and scenes. Given what we learned in the first part of this book, this is just what would be expected. Our understanding of the size and shape of spaces is based on what can be taken in at a glimpse, and it begins to falter when it must find ways to incorporate the unseen features lying behind open and closed contours.

In addition to simple size, isovists have many other properties. We can define isovists in terms of their jaggedness (isovists that have very long borders compared to their overall area are more jagged), complexity (isovists that have many corners are more complex), symmetry (the number of axes of symmetry that they contain), and in many other ways, limited only by the imagination of the researcher. Basic isovist properties such as these can then be combined to characterize more complex properties of a seen space, such as its apparent spaciousness, openness (a combination of jaggedness with some other isovist properties), complexity, and order (a combination of symmetry and some other properties indicative of redundancy).

Do these kinds of isovist properties influence our preferences for spaces? Does the comfortable little corner that you remember using as a perching spot at the family cottage or in your grandmother's farmhouse hold such attraction to you because of the shape of the space that encloses your view? Some experiments suggest that this is a good possibility. When people were shown simulations of spaces with distinctive isovists on large screens, they were able to rank them in terms of their complexity, pleasantness, and interest, and many of their rankings showed impressive relationships with the measurements of the spatial isovists. We find spaces with high complexity and symmetry to be most pleasing, open and symmetrical spaces to be most beautiful, and open and complex spaces to be most interesting.[3]

Given such findings, it isn't at all surprising that we tend to spend more time in particular locations in our homes. Findings such as these may also go some way to explaining the success of fluffers in heightening the interest of potential house buyers. It isn't very likely that such professionals would resort to the extreme of moving walls, but placing furniture in spots with favorable isovists might help to draw visitors to the best locations within a space and then imagine themselves sitting in their own cozy armchairs with grand vistas, symmetrical and jagged isovists, complex and interesting spaces. We seem to have a deep, intuitive understanding of how isovists influence feeling, so it would not be surprising if a group of design professionals had managed to train themselves to heighten their sensitivities to such influences without having had any formal training in architectural theory or isovist analysis. Staging divas are probably good intuitive-cognitive scientists.

THE EVOLUTION OF COMFORT

We are beginning to make some progress in understanding what kinds of spatial arrangements appeal to people in their homes, but it would be of great interest to know about the origins of such emotional attachments to particular places. Much thought on such questions has revolved around people's attachments to natural places, something we will deal with more extensively in chapter 11, on greenspaces. But it is likely that the same kinds of mental processes that guide our preference for certain arrangements of nature also operate inside buildings.

Jay Appleton, a geographer by training, tried to draw a connection between our preference for particular types of visual scenes in nature and our evolutionary history. Was it possible that the basis of our aesthetic preference for certain types of views was biological? Do we prefer to look at certain shapes of space because our evolutionary forebears, naked apes struggling for survival on the African savannah, would have found survival benefits in placing themselves in positions in space from which such views were possible? The cornerstone of Appleton's argument has come to be known as the duality of prospect and refuge.[4] We prefer to be in positions that give us some visual cover (refuge) but from which we can look out over large vistas of space (prospect).

Though the simple idea of prospect and refuge might not be able to account for all the findings related to preferences for isovist size, shape, and complexity, it might certainly account for some important elements of it. At a pragmatic level, it might be argued that my preference for the chair in the middle of a large room is related to my desire to keep tabs on my busy family and to know who has entered and left my house through the side door. But this doesn't explain my pleasure and ease in that same chair when the house is empty or when all the children are long in their beds. Appleton's suggestion is that the preference for long views from sheltered positions (in my case, the corner of a room) is perhaps written into our DNA from days when our overwhelming daily concern was the location of the closest saber-toothed tiger rather than the nearest Starbucks.

In her insightful book on the evolution of house design, *House Thinking*, Winnifred Gallagher suggests that some successful architects have a strong intuitive sense of the power of prospect and refuge.[5] Frank Lloyd Wright, for example, was fond of building alcoves with low ceilings, especially in cozy spots near the hearth. In terms of primitive survival, a sheltered spot near the fireplace must be considered as the archetypal refuge from which to look out on the grand prospects offered up by the rest of the house. Similarly, Christopher Alexander, in *A Pattern Language*, his encyclopaedic recipe book for successful city, town, neighborhood, and house planning, suggests varied ceiling heights as an important design principle, particularly in areas of quiet repose such as alcoves in bedrooms designed to contain beds.[6] In the same way that house stagers may have some intuition for what attracts buyers to a property, good designers and architects have strong sensibilities for how the aesthetics of space can contribute to successful and comfortable abodes. Theoreticians like Appleton lend a scientific rationale to principles that long experience has demonstrated to be effective.

Before going on to consider some other aspects of the influence of spatial cognition on our behavior and preference for certain types of houses, it would be wise for us to take a step backward to make sure that we understand what houses are for. If we use the same approach as in earlier chapters, we might begin by comparing our own homes with those of other animals. Surprisingly, our closest living relatives, the other primates, make little effort to build homes. Those animals that do build homes do so for the simplest and most obvious of reasons, it seems—to provide protection from the elements and from predators, and perhaps to provide a physical base from which to nurture young.

How different are human homes? Why do we live in houses at all? For a resident of a temperate climate like that found in much of North America and northern Europe, the answers seem obvious. The main functions of a house are to provide an insulating shell to protect us from the environment and to provide us with a safe repository for our possessions.

Amos Rapoport, a pioneering anthropologist who helped to define the field of environmental psychology, argued persuasively that practical considerations must be only a part of the answer.[7] If one looks at the range of house types produced by various cultures, there is only weak evidence of a correlation with either climate or the availability of different types of building materials. Primitive peoples in tropical climates didn't necessarily build simple dwellings that promoted coolness. On the contrary, some very elaborate and climatically stifling buildings were typical of people who lived in hot climates. One good example is the elaborate system of dwellings used in parts of the South Pacific, in which separate and sometimes closed-in buildings were constructed for men, women, and for eating. In contrast, some cold-weather dwellers lived in simple huts. Rapoport argued that the shape of the house one lived in spoke at least as strongly about one's beliefs, values, and culture as it did about the bare necessities of survival. Round houses, for instance, are very economical. They can be easy to build and adaptable to a wide range of building materials. In spite of this, round houses are rare. One reason is that most cultures have placed a high value on the orientation of the rooms in a house with respect to the larger site and to the other buildings in the neighborhood.

This tendency to carefully orient built spaces reaches its apotheosis with the Asian practice of feng shui.[8] There are many schools of feng shui. In some quarters, feng shui might be considered a rather mystical and New Age discipline, somewhat on a par with astrology and other forms of divination. Practitioners of what is sometimes called "McFengshui" in North America, by claiming to be able to promote wealth, harmony, and successful marriage by pointing one's bed and toilet in the right directions, have no doubt contributed to such opinions. Nevertheless, serious schools of feng shui, with theoretical roots that are thousands of years old, can include a comprehensive effort to align one's home with prevailing geological forces such as magnetic fields, and principles that guide the construction of well-organized and connected interior spaces with respect to the world outside the walls of the house. Adhering to feng shui principles or other cultural practices that connect our homes to the world outside, both natural and supernatural, can be difficult in houses that lack the contours provided by square or rectangular rooms.

Another good example of the influence of culture on the built environment comes from the prevalence of courtyard homes in certain parts of the world, particularly in Islamic countries. One of the benefits of a courtyard construction is that it affords some privacy for residents of the space, but, within the courtyard, it also allows the construction of separate buildings for men, women, and the generations within a family. In this way, courtyards enhance the privacy of the larger family unit and give them shared social spaces away from the public eye, but the separate buildings also allow physical demarcations of family hierarchies within a single courtyard.

Rapoport's life work consisted of an extended effort to demonstrate that for those of us who build our own homes, the principles that determine how we put together our lived spaces transcend the simple need for shelter and protection from enemies. Our homes are outward manifestations of our beliefs, desires, and perhaps our deep fears as manifested in our own culture.

What about modern homes in North America, Europe, or other parts of the world? Unlike in the cultures or times of which Rapoport speaks, few of us have built our own houses. Rather, we select from those that are offered to us by developers or real estate agents, and we may have only minimal input into the shape of the spaces we occupy. The irony that Rapoport identifies is that at a time when there is greater variety in available building materials than at any other time in human history, and when many of us have fewer economic constraints than those whose houses he describes, the configurations of human dwellings appear to show less variability than at any other time in our past. One reason for this is certainly the great shift from a time when most of us built for ourselves to a time when the majority of dwellings are designed and built by specialists who may be more concerned about their bottom line than about our comfort and enjoyment of our homes. Another reason identified by Rapoport is the diminution in importance of higher-level cultural constraints on home architecture. Few of us worry seriously about how our house is oriented with respect to the coordinates of sacred space (though we may covet a southern exposure for our garden), and we are more concerned with the number of bathrooms than with the orientation of the toilet with respect to the bed in the master bedroom.

Aside from cultural and economic considerations, technological influences have also drawn the attention of designers away from the construction of habitable dwellings. In a recent conversation I had with Robert Jan van Pelt, an architect with a strong interest in the history of ideas, he argued that most designers and architects today focus upon either the very large—airports, museums, and city halls—or the very small—corkscrews, lemon squeezers, and chairs. The large, one-of-a-kind architectural creation like I. M. Pei's Louvre Pyramid or Frank Gehry's Guggenheim Museum Bilbao can serve as a long-standing physical monument to the ideas of the architect that is talked about, photographed endlessly, and seen by all from afar. The small household object, sometimes crafted with the same exquisite concern for detail as a large building, can be reproduced in vast numbers using modern methods of mass production. This is not only profitable but also serves as a different route by which the ideas of a designer can penetrate deeply, be seen by many, and influence much of our behavior.

The ordinary dwellings in which we live can lack the glamour of the huge hotel or office tower and the promiscuity of the kitchen utensil. For the designer, there is a bit less appeal, and so for the everyday consumer, there is less variety.

In spite of the stultifying sameness of our cookie-cutter housing developments, it is possible to identify certain patterns that do connect with modern culture and perhaps psychology as well, but one must look closely to see them. One of the most important functions of the inner design of house spaces is to regulate contact not just between members of the household but also between the residents of a space and outsiders. We already saw this at play in the use of courtyard houses to keep outsiders beyond the threshold as well as to make allowances for regulated social mixing of hierarchies within the household. In the famous hutong neighborhoods of Beijing, to use another example, a few splendid courtyard houses remain intact. To the passerby in the street, the outer ramparts of these courtyards can display nothing more than a bland brick wall with a humble door. The inner courtyard, though, can contain appealing landscaping and complex building forms with specialized functions for different members of the extended family.

In North American architecture, interesting cultural distinctions can be revealed in the way that the transition from public to private space is managed. Though this is somewhat less true now than it was twenty years ago, it is still largely the case that suburban front yards set houses well back from the street on unfenced lawns. This sets up what amounts to a semiprivate space in front of the house. In English and Australian suburbs, on the other hand, one is more likely to see a sharper distinction between public and private spaces in which fences clearly demarcate front gardens as private spaces. Though these fences may be low enough to be merely symbolic, their meaning is still clear.

We can also see the evidence of cultural change in North American houses. The traditional house plan in North America still often includes a formal dining room, for instance, even though this room, normally occupying pride of place in a space adjacent to the kitchen, is seldom used. Few people I know who have a formal dining room use it with any regularity, unless they lack seating space in their kitchen. Many people convert such spaces to other uses, but as these rooms were designed for seated dining, they are not always optimally placed for other uses. One good example of this is my reading armchair location in my house. The chair, housed in a room originally designed as a dining room, does not see as much use as it might have in a space designed explicitly as a comfortable and well-placed reading room.

In times past, one of the most important architectural elements of a dwelling was the foyer. As Winnifred Gallagher points out in *House Thinking,* the main function of the threshold is to manage the transition from the outside world to the inner space of the house. In this light, we can see the foyer as the beginning of a kind of stage performance, the opening act of a drama in which the curtain opens to reveal the detailed inner life of the homeowner. Grand homes of the past included stunningly ornate entry rooms that clearly foreshadowed the character of the house that lay beyond. In modern homes, the foyer is usually designed as a more informal space, often not much more than a coat closet beside a door. In the worst cases (such as the house I own now) the foyer can be missing entirely.

One exception to the simple, informal foyer can be seen in certain grandiose designs in the outer suburbs, the so-called upscale executive designs. In such homes, entrance foyers can consist of multistory spaces complete with overlooking balconies and grand chandeliers. As Gallagher points out, the effect of such entryways can be psychologically negative, causing visitors to jerk their heads upward in anxiety as they walk through the front door, as if they have found themselves at the bottom of a mineshaft. The irony of such grand foyers is that they are seldom used, as the majority of owners of these houses drive directly into attached garages and enter through humble back doors into laundry or mud rooms. It often seems as though the main function of the foyer, as the part of the house that makes that important first impression, is more to stun potential buyers into submission than it is to exert any kind of positive influence on the owners of the house or its visitors.

One probable effect of such foyers is to increase the insular separation between the outside world and the inner world. Like an airlock protecting our spatial mind from the intermingling of the great outdoors and our pristine inner spaces, the foyer takes advantage of a spatial mind predisposed to cleave the world according to what can or cannot be seen. This is an issue that we will explore in a later chapter when we look at the influence of spatial cognition on our relationship with the natural world. There can be little doubt that the modern foyer, along with many other inventions of modern architecture and design, has helped to sever our relationship with nature. If we are to find a way to survive on this planet, we will need to find ways to heal this enormous rift.

HERMANN MUTHESIUS AND THE ENGLISH HOUSE

What I have said so far suggests that the size, shape, and configuration of the spaces inside our houses can have an influence on how we feel and where we go for comfort, for solitude, or for conversation. Some of the general principles underlying such behavior have been uncovered by architects and designers using largely intuitive experiential methods.

Christopher Alexander is perhaps the best known of architects who have tried to apply a thoroughgoing understanding of the interactions between people and spaces in order to design more functional dwelling spaces. A brilliant polymath, Alexander received the first Ph.D. ever awarded in architecture by Harvard University while simultaneously working in the fields of computer science, transportation theory, and cognitive science at both Harvard and MIT. (His thesis, published as *Notes on the Synthesis of Form*, has been required reading for students of computer science for many years.)[9] Alexander was recruited by the University of California at Berkeley in 1963, where he has remained ever since.

In a series of books culminating in the four-volume work ambitiously entitled *The Nature of Order*, Alexander argues for the integral connections of everything from quantum mechanics to living rooms to religious epiphany. What links all these things together, he says, is a set of principles that describe the ways in which what he calls centers, which are explicitly spatial, contribute to "wholeness." According to Alexander, these rules govern everything from how the large-scale matter of the universe is ordered to what size and shape of living room in a house makes for peace and stillness.[10]

Given my thumbnail sketch, you might be left thinking that Alexander is something of a New Age mystic, but such an impression would be deeply misguided. Alexander deals with the notion of centers and wholeness at a completely pragmatic level, using it to explain such things as how the different parts of a well-made chisel (its blade, handle, connections) contribute to its beautifully simple functionality. In almost the same breath, Alexander draws an analogy with the connection between the length and orientation of a hallway and the quiet dwelling space it is attached to, making it clear how the two parts (two centers) contribute to the organic unity of the whole. Alexander's main aim is to find a way to explain how the geometric properties of spaces have "the power to touch the human heart."

Though his agenda is sweeping, his belief that space influences feeling and movement at a deep level is very much in accord with what we are now beginning to document with scientific experiments. The underlying reasons for these types of influences are buried in the nature and organization of the parts of our mind that have evolved to contend with problems of physical space. Alexander may not mention isovist analysis explicitly, yet it seems certain that some of the ways the shape of space influences behavior can be explained using the same analyses that I described earlier, and, in turn, that these analyses work so well to predict feeling and movement because of the size and shape of our spatial brain.

Sarah Susanka, the successful architect and popular author of the groundbreaking *The Not So Big House*, bemoans the North American tendency to equate square footage with happiness. She has demonstrated through more than two decades of building, writing, and case studies some of the same principles espoused by Alexander. Susanka insists that it is the quality of space, rather than its quantity, that influences our behavior. How we are attracted to spaces and how we thrive in our own dwellings has much more to do with the configuration of spaces and the small finishing touches, such as alcoves, built-in furnishings, and the quality of light, than it does with pure geometric horsepower: the number of tape measures needed to measure our expanses of real estate. What Alexander has managed to convey over a life's work of practice and writing, Susanka has attempted to boil down to a somewhat simpler level, allowing more of us to take advantage of what is known about how shape influences feeling.[11]

Hermann Muthesius, though not exactly a household name, is well known to students of architecture. Muthesius was recruited by the German government in 1896 to work in England as a cultural attaché. Rumors have swirled from time to time suggesting that Muthesius had a secret role as a spy (it does seem to be true both that he was personally appointed by the Kaiser and that part of his work consisted of careful documentation of English infrastructure, including railroads and heavy industries). Yet mostly he occupied himself scrutinizing the architecture, furniture, and manners of the English household. He quickly "went native," packed an inordinate amount of traveling into a short tenure in England, and studied every detail of English life, from the habits of afternoon visitors to the placement of soap dishes near bathtubs. The culmination of this work was the monumental three-volume *The English*

House and, in it, the suggestion that an integral connection existed between the design of the English home and what Muthesius saw as their enviably successful way of life.[12]

For example, Muthesius was much struck by the English lack of ostentation in their dealings with houseguests. The common German practice consisted of carefully stage-managing visitors such that private quarters were kept hidden away and that the best, biggest, and most formal and impressive rooms and furnishings were highlighted. The English seemed to simply invite outsiders into the most intimate corners of the house:

> *It is amiably taken for granted that no special arrangements will be made for the visitor. He is one of the family and can do or not do as he wishes, like any of the others ... Everything goes on as usual and the visitor is spared the embarrassing feeling—that ultimately obliges him to leave—that he is upsetting the routine of the house ... True courtesy lies in the very absence of conspicuous marks of it.[13]*

Muthesius noted that the absence of physical separation between guests and residents in English homes was brought about by the artful arrangement of spaces. For one thing, the rooms of the English homes that Muthesius studied possessed what he called "very good wall spaces." What he meant by this was that typical rooms in these houses were entered from hallways and were almost always cloistered behind doors. Door hinges were arranged so that the doors always opened inwardly upon the private spaces behind them. In this way, someone entering the room would begin their experience of the room by first seeing a tiny crack of an opening before seeing the entire room. This way of hinging doors produces an isovist that increases in size slowly and methodically, much as Muthesius saw the English style of life. This method of door hinging would have the further advantage that anyone already inside the room would have ample warning of the entry of the visitor, and so have time to prepare for the visit.

Muthesius contrasted such arrangements as he saw in English houses with German houses of the time, in which one room commonly contained an open doorway to another room, making the whole affair more like a succession of ostentatious entry halls to the living space rather than a comfortable set of quiet, contemplative spaces shared by guest and owner alike. More generally, he saw the German approach to the design of dwelling spaces to be showy, wasteful, and stiff compared to the more relaxed, humble, and disingenuous approach espoused by English architecture.

Arguments such as those of Muthesius are interesting because they suggest that there are interesting relationships between the way we design built spaces, the way that such spaces interact with our psychological makeup (the manner in which the organization of space makes or breaks intimacy, for example), and our cultures. Not only can the architecture of house spaces be used to reinforce cultural norms but it may also amplify them. Muthesius saw in microcosm in the organization of the English home many of the reasons for the apparent economic and social success of England at a time when Germany was struggling.

As we have already seen, such clashes of culture and space may have exerted enough of an influence over human behavior to cause bloodshed in our own times. Mohammed Atta, incensed by the imposition of Western architectural ideals on Muslim streetscapes, was ultimately prepared to sentence thousands of innocent people to death to exact payment for the insult. Later in our story, when we look at city space, we will see another example of a head-on collision between the organization of built spaces and the cultural values of those sentenced to live in them, a clash that contributed to death and destruction on a grand scale. Before we venture into the city, though, we must spend some time with the larger interior spaces where we work and play.

CHAPTER 8
WORKING SPACE

HOW THE GEOGRAPHY OF OUR MIND INFLUENCES
OUR HABITS OF WORK AND PLAY

If a train station is where the train stops,
then what's a work station?
ANONYMOUS

St. Peter's Basilica, with a capacity of 60,000 people, not only is the largest Christian church in the world but, as a landmark punctuating the Roman skyline, it resides in the same pantheon of immortal city identifiers as the Eiffel Tower, Big Ben, or Rio de Janeiro's mountaintop statue of Christ the Redeemer.

St. Peter's is much more than a landmark, though. It is an icon of the Roman Catholic Church, the venue for religious events presided over by the Pope, and the reputed burial place of the biblical St. Peter. In other words, this magnificent structure stands as one of the most important pieces of religious architecture in the world. It is a sacred space of the highest order.

When I visited the basilica for the first time, I joined the long lineup of visitors sweltering in the hot July sun. As we made our way toward the main gate, we endured the inspections of officers of the Vatican Guard who ensured both that we were not carrying contraband and also that we were dressed appropriately. This only served to enhance our excitement about entering one of the architectural marvels of the world.

I will never forget the feeling that engulfed me once I had cleared the doorway and the gigantic space was in full view. It felt as though the wind had been knocked out of me. I was frozen to the spot for the better part of a minute as my eyes tried to make sense of both the sheer scope of the space and its overwhelming kaleidoscope of textures and colors. When I was finally able to move, I found myself walking slowly and carefully, clinging as much as possible to the sides of the building like a timid mouse in a lion's den. As I looked around, with a sudden and inexplicable lump in my throat, I noticed that many other visitors seemed similarly affected. Most people were quiet and reverential. They took tiny steps across the threshold. One man actually slumped to his knees. To some extent, these reactions might be attributed to the sacred significance of the site. Most visitors, even summer tourists, would realize that this magnificent structure represents the spiritual center of the modern Roman Catholic Church. Not only this, but the site is stuffed with incredible numbers of precious artifacts, holy relics, and stunning works of religious art. This alone, one might argue, would be enough to render many people a little weak in the knees.

In addition to the overwhelming scenery, though, I would argue that impressive spaces such as those found in huge churches and cathedrals are designed explicitly to influence our feelings, our moods, and even our patterns of movement. The same spatial influences that might cause us to gravitate to a favorite chair in a humble dwelling can be amplified to hit us like a sledgehammer in a differently organized space. This is demonstrated convincingly in a sacred setting, but just the same principles that might choreograph the high drama under a cathedral roof are also at play in the more mundane spaces of our everyday lives.

Our offices, shopping malls, government buildings, and casinos all have design elements that help to control how we move, where we pause and rest, and what feelings we have along the way. In the best cases the principles of spatial design can enhance the function of a building. A well-designed office building can make us happier at our jobs, and an artful casino can encourage us to gamble away too much of our money. Conversely, poorly designed spaces can work against our aims. Government buildings in which we lose our way quickly will frustrate us, and institutional buildings that cut us off from the outside world make us angry. In this chapter, we will focus on what I've called working spaces, and my intended meaning is a bit of a double entendre. Though we will look at the design of spaces in which we work at our jobs, our more general interest is in how the spaces inside larger buildings can "work" us by sculpting our patterns of movements while we are inside them.

FINDING OUR WAY THROUGH BUILDINGS

When Bob Propst invented the Action Office concept in 1965, he unwittingly (and much to his later horror) helped to set in motion developments in the organization of workspaces that would culminate in an epidemic of unhappy and unhealthy office workers trapped in stultifying beehives of cubicles.[1]

Propst's invention of the Action Office, a modular design system in which "panels" could be matched with function and combined like so many building blocks to produce a highly functional workspace, was meant to begin a revolution in office design. Propst's aim was to bring the design of workspaces into closer accord with what he saw as revolutionary changes that were taking place in the world of work during the 1960s. For one thing, workers were being required to deal with huge amounts of information necessitating complicated workflow and communication requirements like nothing that had been seen before. The Action Office was designed to facilitate these new requirements in environments that were spacious, comfortable, attractive, and ergonomically sound. Indeed, the Herman Miller Company, with a $1.5 billion market share as one of the world's leading providers of office furniture and environments, continues to offer versions of the Action Office based on some of Propst's original ideas.

Somewhere along the way, though, much of Propst's thinking was hijacked. In place of Action Offices designed to provide thoughtful integration of the needs of both an individual worker and collaborative groups, there arose the nefarious cubicle, essentially designed to maximize the number of workers who could be housed in expensive urban real estate but not always with adequate consideration given to how the shapes of spaces made by the great nests of cubicles might affect navigation, communication, or the general state of mind of workers. In contrast to Propst's Action Office concept, classic cubicles isolate workers in any one of a vast number of small, identical workspaces while effectively cutting them off from co-workers with the use of high partitions.

There is no shortage of satire dealing with the problems of cubicle culture, from Dilbert cartoons to the cult movie classic *Office Space*. This form of office organization has also given rise to lexical novelties. "Gophering" is exactly what it sounds like—the practice of standing up to raise one's eyes above cubicle walls to take in a larger vista. Gophering in cubicle farms is elicited by exactly the same kinds of events that produce this behavior in the animals from which the name is taken—a desire to take in additional information in the face of some kind of threat or instability such as a loud noise (a shouting co-worker or supervisor) or some other kind of stimulus such as an unusual smell (of food, hopefully). Though it sounds funny, gophering is a genuine response to an environment that has spatial limitations. People are peering over the tops of cubicle walls in part to make up for an information deficit that has been produced by the configuration of space in their office.

Those of us who have to work in cubicle farms may find them soul destroying, and those of us who don't might be amused by them, but the main thing is that our behavior in such work environments points up the importance of the configuration of space not just in our homes but in the larger indoor spaces of our lives—spaces in which we work, play, or educate or entertain ourselves. Just as the right kind of house can make us feel happy, thoughtful, excited, or creative, so can workplaces and other, more public spaces influence our feelings and behavior in important ways, both positive and negative.

When thinking about the spaces inside our dwellings, we are often preoccupied with the influence of space on repose. Where are the best resting or thinking places? Where do we bring company to sit? In larger buildings such as offices, schools, courthouses, government buildings, and shopping malls, we are likely to spend more of our time in motion, and the way that we move from one place to another is likely to influence the quality of our experiences or the efficiency of our workday. To understand how the organization of physical space affects our movements, we need to look beyond isovists to see how the appearance of inner spaces is influenced by movement.

Researchers interested in the technical properties of space and how those properties change for us as we move have defined new measures that are related to the isovist analyses we considered in the last chapter.

One such measure is called a visibility graph. To understand how visibility graphs are constructed, remember that the isovist represents all visible locations, as defined by both open and closed contours, from a single position in an interior such as the large family room/kitchen in my house. Isovists work well to define all that can be seen from a single point in space, but to understand how the shape of space influences movement, we need to consider the connections between isovists.

The isovist that is available to me changes as I walk across a room, from one side to the other. A visibility graph is a symbolic way of representing such changes using the concept of intervisibility. Two locations in a space are said to be intervisible if an observer standing at one of the two points could see the other. Imagine a complex space, such as an irregularly shaped room in an art gallery, filled with an orderly grid of points, where each point represented a potential viewing position. At each viewing position, some of the other points would be visible and others would not. A visibility graph for this space would show all of the intervisible points. This type of representation is interesting because it shows how our perceptions of space change as we walk about. Just as an isovist can be used to characterize the size and shape of a piece of space from a single viewpoint, a visibility graph can be used to do much the same kind of thing, except that it reflects the way that the appearance of the space changes with our movements.[2]

For example, the "stability" of a space is a measure of how the number of intervisible locations varies in different parts of a space. A bland, rectangular space with no visual occlusions, such as a large great room in a modern suburban home, would be a very stable space. A more jagged arrangement with lots of walls and barriers jutting out, a spiky space in other words, would be much less stable. Another spatial measure that can be derived from the visibility graph is something called the mean shortest path length. To calculate this, we measure the shortest distance from each point in our grid to every other point and then we calculate the average. This value will depend very much on the overall shape of the space as well as on any barriers to movement that might be within it, such as pieces of furniture in a home or desks in a workspace. A measure like mean shortest path length is different from isovist measures because it reflects not only the shape of a space but also its possibilities for movement. I may be able to see the window from behind the outer edge of my cubicle wall, but if I decide to walk to the window I need to walk around another row of cubicles.

If visibility graphs were just another cool toy for mathematicians interested in space, they would not be worth our trouble here, but there are intriguing indications that such graphs, and many other related tools for analyzing space, can make surprisingly accurate predictions about how we move through and spend our time in a complex configuration of space. The Space Syntax Laboratory, a part of the Bartlett School of Planning at University College, London, has had marked success in predicting how people move through spaces on the basis of the graphical tools I have been describing.[3] Because most of the work of this group is concerned with the influence of spatial configuration in larger urban settings, we will deal with it more extensively in the next chapter, on city space, but many of the principles used to steer urban planning apply equally well to interior spaces.

For example, an analysis of intervisibility and shortest path length values for the Tate Gallery in London has been used successfully to predict where visitors will congregate in the gallery. The Space Syntax Laboratory has used these kinds of analyses to advise the gallery on the effective placement of exhibits to encourage the flow of people and avoid pedestrian gridlock. What is most remarkable about the success of these analyses is that they work well even when little or no account is taken of what kinds of objects will actually be in the space. Analyses of space can be based on the raw configuration of space—its shape rather than its contents.

Predicting where people will congregate in a space based on its shape can be a useful tool for planners. A gallery owner wishing to draw maximal attention to a particular work of art could use visibility graphs to determine where best to place the work. A designer of shopping malls could engineer a space so as to steer people to some locations and away from others. A committee of workers trying to design an efficient workspace could use a basic understanding of space to facilitate a particular group dynamic by engineering the manner in which people interact in the space. Sometimes, such strategies can be explicit and obvious and don't require any mathematical measures at all. For example, most people are aware of explicit spatial strategies used by grocery stores to ensure maximal traffic, such as placing the dairy case as far as possible from the entry door so that customers dashing in for a carton of milk must navigate many aisles of products they didn't set out to buy. In other cases, much more subtle methods can be used. To understand a few more of these subtleties, we must take our analyses of the shapes of space just a little further.

PREDICTING WHERE WE WILL GO

Apart from making predictions about how people will move through space and where they might congregate, mathematical analyses of the shape of space can help us to understand how well we can find our way around a building. We all know that some buildings seem intrinsically more difficult to navigate than others, but it isn't always clear why this might be.

As legend has it, the building where I work, the psychology building at the University of Waterloo, in Ontario, was designed so that its shape corresponds roughly to the shape of a brain. Many visitors or even longtime students complain that they have difficulty finding their way about because of the lack of distinguishable landmarks in the corridors. Each hallway is the same size and shape, and the whole building can seem like a beehive of identical orange office and laboratory doors. Other reasons for the poor navigability of my building have to do with the way that spaces are connected. The connections between spaces can be captured using a slightly different technique, referred to as space syntax. In space syntax analysis, we try to use a simple graphical method to describe the way that different regions of space are connected to one another, just as we might use the linguistic form of syntax to understand how a sentence is constructed.

To begin, we draw a diagram in which each room is reduced to a single point, and then draw lines connecting all the points that are directly accessible to one another, a "point-and-stick" representation. Such lines would most commonly represent hallways, but when two rooms are directly adjacent to one another, the doorway between the rooms could also be represented by a line. The diagrams below show an example, using the ground floor of my own house. Figure 8 shows the actual layout of the rooms, and Figure 9 shows the floorplan reduced to points and lines.

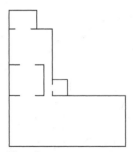

Figure 8: The ground floor of my house, shown as a standard floor plan

Figure 9: The ground floor of my house, shown using space syntax analysis

If we formalize space in this manner we can obtain some simple measures by calculating things like the average number of steps (where a step is a hop from one point to another, connected point)

required to get from anywhere in the space to anywhere else in the space. Diagrams such as Figures 8 and 9 are so simple it would hardly seem worthwhile to carry out these kinds of arithmetic operations, but for more complex spaces in larger buildings, such analyses can be revealing.

One such measure, referred to as intelligibility, characterizes the degree to which the shape of any small part of a space reflects the shape of the whole space. Think of this as a kind of correlation between the spatial characteristics of the whole building and the characteristics of any small part of the space. An intelligible building is one in which the hallways that one needs to use most often to get from one place to another are also the ones that intersect with many other hallways. It isn't hard to imagine an unintelligible building: it could be one in which a hallway intersecting many other hallways leads exactly nowhere, or one in which a small area with very few connections must be navigated to get almost anywhere else in the building. A building that contained a regular grid of hallways would also be considered unintelligible because all hallways would appear to be more or less equivalent. They would all present a similar appearance and they would all be equally connected to one another. Certain types of spatial puzzles, such as hedgerow mazes, are often designed explicitly to have very low intelligibility. As was the case with visibility graphs, the marvelous thing about intelligibility in this formal sense is that it correlates very well with behavior. People get lost in spatially unintelligible spaces much more often than they do in intelligible ones.

THE ARCHITECTURE OF SPACE AND THE SHAPE OF THE HUMAN MIND

It might seem strange at first that these simple diagrammatic representations of space, in which shape and volume are reduced to nothing more than dots and sticks, work so well to predict our movements in a building. The success of space syntax has to do with the manner in which our mind deals with problems of space and navigation, as we saw in the first half of this book. For one thing, many types of space syntax analysis completely disregard the metrics of space. Simple diagrams of rooms and hallways like Figure 8 collapse all information about the sizes of the rooms that are represented by the dots, or the lengths of the hallways that are represented by the lines, yet they can make highly accurate predictions about how people will explore the spaces and how well they will be able to locate themselves. This flagrant disregard for size and shape should remind you of some of Barbara Tversky's findings regarding the schematization of space. By asking research participants a series of simple questions about geography, Tversky was able to show that we straighten curves, simplify geometry, and reduce complex spaces to a simple series of points and lines. Just as our mind represents space as a kind of dimensionless topology, formal models of that topology can predict our movements through space with surprising accuracy.

What is perhaps even more surprising than our apparent disregard of distance when wayfinding is that our behavior inside buildings can be simulated to a high degree of accuracy by replacing us with "agents"—simple bits of computer code that are designed to behave according to a small number of rules, the most important of which is "always move in the direction that is the most 'open.'" If a computer simulation is drawn to approximate the layout of a real building, and a few such agents are unleashed on the simulation, like tiny Pac-Men that chomp down the hallways guided only by simple rules, the agents have problems where humans would get lost and spend the most time where humans would spend the most time in the real building.[4] This finding draws a direct line from the point-and-stick representations of the syntax of built space, through our ability to quickly recognize the size and shape of spatial isovists, to our specialized brain, heavily biased toward the visual sense and an understanding of layout, landscape, and vista but less sophisticated when it comes to under-standing the connections between the seen and the unseen.

In addition to intelligibility in the formal sense, some other factors influence the wayfinding friendliness of a building. One of the most complicated building designs, from the wayfinding perspective, is one that includes wings or hallways that intersect one another at oblique angles. Remember that because our minds are always looking for ways to simplify mental models of space, we have a tendency to align differ-ent regions, straighten curves, and smooth out jagged edges.

If you were to look down from above on the brain-shaped building where I work, you would see something like a doughnut-shaped structure with a ring of offices and laboratories organized around a central courtyard. Quite apart from the larger symbolism of the building, this is, in some ways, a lovely architectural idea. For one thing, it maximizes the number of offices that have win-dows, which can help to connect the building's occupants to the outside world. For the wayfinder, though, the inner hallways can, quite literally, pose some treacherous curves. When I first took up residence in the building, I slowly learned the way from my office to various other locations around the ring, but my mental repre-sentation of the space was always as if the main central hallway consisted of one straight line, rather than a ring that circumnavi-gated the building. Using a simple route-based strategy, I learned that some locations required a left turn from my office door and others required a right (though truthfully, because the hallway was a ring, turns in either direction would eventually lead me to any destination on the floor). On one occasion, I followed a colleague from my office to his lab, but as he walked out the door he turned the "wrong" way. When I called out to correct him, he looked over his shoulder, eyes twinkling and eyebrow arched, and gestured for me to follow. When I arrived at his lab, having taken a route that went against the grain of habit, nothing felt right. I was vaguely disoriented, as if I'd arrived at his lab through a peculiar wormhole in the fabric of space-time.

Although many architects are aware of the principles that underlie successful wayfinding in buildings, these principles must sometimes take a back seat to other architectural concerns, such as economics or even aesthetics. At a time when many cities are looking for architects who will design signature buildings that will produce recognizable landmarks or even attract curious tourists, the pragmatics of designing a building in which people do not become lost easily can be a minor consideration. Indeed, some of the most dramatic architectural creations in recent years, filled with either sinuous organic curves or the sharp angles of oblique, crystalline forms, though they make distinctive contributions to city skylines, do not admit of easy wayfinding. In such cases, the remedy is often to help to steer occupants of the building using carefully crafted signs and graphic aids such as "you are here" maps. These kinds of landmarks and spatial crutches can work well to remedy the psychological flaws of poorly designed configurations of space, and several companies specialize in crafting such navigational support systems for buildings, especially in the health care sector, where episodes of disorientation by patients or visitors could be stressful or even life threatening.

———————

Understanding how our spatial cognition influences how we move and where we rest can often be used to exert a kind of social control. Many examples of this use of spatial design principles can be found in commercial buildings such as supermarkets, department stores, and shopping malls.[5] In department stores, different sections can be placed as if to set the stage for a kind of story in which the shopper plays the starring role. Cosmetics are placed carefully near other adornments, such as jewelry and purses. Men's sportswear is kept respectfully apart from the tiny black dresses women wear to the fanciest parties.

The placement and design of food courts are also carefully managed to exert control over behavior. Unlike department stores, where mall owners hope that customers will linger for as long as possible with wallets in hand, food courts are designed to discourage lingering. Such areas are usually very open. Enclosing walls, and the refuge they offer, are avoided by arranging wide aisles around the outside of the seating area that are designed to draw people to the service counters. Food courts are brightly lit, often with skylights and high ceilings. Tables are arranged in such a way as to discourage groups of diners any larger than two. The effect, very much like trying to have lunch in the middle of an overdone foyer in a suburban McMansion, is artfully contrived to encourage people to slap down their money, wolf down their food, and plunge themselves back into the shopping fray.[6] Perhaps a more apt metaphor would be to imagine primitive *Homo sapiens* sitting down for a nice lunch in the middle of a wide open stretch of savannah. He would undoubtedly run a great risk of *becoming* lunch, rather than *consuming* it, so would be unlikely to linger for dessert.

Shoppers might be corralled out of food courts and into high-end jewelry sections by the subtle manipulations of space, but other contexts where the explicit use of the size and shape of space to exert social control on our behavior are even more extreme. In the gigantic gambling palaces of Las Vegas or Monte Carlo, shrewd designers understand that the placement of each hallway, crap table, or slot machine can influence the amount of money taken in by the casino.

Currently, there are two main theories about the best way to organize the space inside a casino in order to more quickly liberate the cash hiding in the wallets of visitors. One influential set of studies, carried out by longtime casino consultant Bill Friedman, emphasizes that the best way to maximize the yield of a casino is to focus the attention of visitors on the gambling equipment itself, especially the slot machines. To encourage this laser-beam focus, Friedman encourages the use of low ceilings, narrow aisles, and tight spaces so that the visitor is surrounded on all sides by the flashing lights and ringing bells of the slots. In addition, Friedman encourages spatial designs that explicitly work against good wayfinding—this is one context in which low spatial intelligibility would be considered a business asset. In general, Friedman's philosophy seems to be one of doing all that can be managed to compel the visitors to spend as much time at the gambling machines as possible and to make it as difficult for them to leave the building as possible.[7]

Not surprisingly, given what we've already learned about how people use space, this approach may help to empty the pockets of gamblers, but it isn't necessarily the most pleasant way to spend an afternoon, evening, or weekend.

Another model of casino design, championed by David Kranes, is based on the notion of a casino as a playground. In contrast to Friedman's approach, which almost seems designed to snare visitors in the way a spider might lure a fly into its web, Kranes's design philosophy is that casinos ought to be places where people not only want to come to have fun but also want to return again and again. Kranes argues that casinos should present large, vaulted spaces with beautiful textures and objects in addition to all the paraphernalia of gambling. Quite apart from the excitement of the games and the risks they involve, we should feel that we are in an inviting, spatially intelligible, and perhaps even restorative environment. As Kranes puts it, "Gambling is a curious activity. We want to relax—and we want our blood to boil ... all at once. Want to be both fully *in* and *out* of control—without contradiction."[8]

So what do the scientists have to say about these different approaches to social control of space in casinos? Some of the best work in this area has been conducted by Karen Finlay's group at the University of Guelph. Supported by the Ontario Problem Gambling Research Centre, one of the main objectives of the group is to understand how context effects can contribute to the tendency of some individuals to spend more money in casinos than they can afford. Is it possible that the very shape of a gambling space can encourage us to give away our next mortgage payment? Finlay's work suggests that it is. By placing volunteers into virtual mock-ups of parts of casinos, or even by simply showing them photos or videos of casino interiors, Finlay tries to duplicate the contexts of actual casinos. She and her team administer psychological tests to the viewers of such materials to assess their feelings, moods, levels of arousal, or sense of restoration. Finlay's research suggests that there are distinct differences between the effects of casino designs inspired by Kranes or by Friedman. But she has also found evidence to suggest that people with certain types of personalities might be more inclined to gamble beyond their limit depending on the context in which they find themselves. Specifically, playground-type casinos appear to be more likely to precipitate risky gambling behavior, especially in individuals whose normal temperament inclines them to be generally difficult to arouse.[9]

Findings such as those of Finlay and her co-workers suggest that the effects of the arrangement of space on our behavior might at times be so strong as to cause us to engage in activities that put our lives and the lives of our families at risk. Given the enormous social costs of problem gambling, such issues deserve our close attention.

MAKING WORKSPACE WORK

In addition to the time we spend in large buildings to shop, for entertainment, or perhaps to interact with government officials, most of us spend many hours in such larger inner spaces because of our occupations. The many ways in which the design and configuration of space can influence worker behavior, productivity, and job satisfaction are both fascinating and complex.

At a basic level, the organization of space can be used to control access and regulate privacy within the workspace. One simple example can be found in many office buildings, where there is a correlation between the position of an executive in the power hierarchy and his or her spatial position in a building. Receptionists, almost by definition, are going to be useful only if they are placed where they will be easily discovered by visitors who are unfamiliar with the building. Executives may want to be cloistered in corner offices with limited access in regions of low spatial integration. A more subtle example of the use of space to regulate privacy can be seen in many health care facilities such as hospitals, nursing homes, and chronic care facilities. Distinctions between public space and space designated for staff members can be indicated with explicit signs and locked doors, but it is also possible to engineer the manner in which people will flow through a building using the principles of point-and-stick spatial analysis or computer-simulated agents. Buildings can be designed to minimize confusion, discourage contact between public visitors and those working behind the scenes, and maintain orderly flow of visitors through a setting. Think of the last time you visited a hospital. Such buildings are filled with rich combinations of areas that are accessible to the public and areas such as examining rooms, surgical suites, and physician lounges and offices that are clearly off limits. Though visitors are sometimes kept out of sensitive areas by lock and key (or code and touchpad), it is surprising how often such private areas are unlocked yet largely left undisturbed. Usually, this is not an accident but rather the result of the careful design of space.

Apart from privacy, the artful (or scientific) design of work-spaces can be used to promote desirable workflow patterns, to enhance contact between particular groups of employees, or, in the argot of modern designers, encourage the creation of sponta-neous "thirdspaces," those areas of spatial convergence, the "water-coolers" where people gather spontaneously to discuss last night's television shows and, hopefully, to exchange ideas.

Economics will always be an important determinant of work-space design. When offices are placed in the standardized foot-plates of expensive urban real estate, the tendency is to pack as many desks as possible into a small space. In conventional cubicle arrangements, workers are sometimes set up in arrangements like Manhattan city blocks, with straight lines, narrow corridors, and an unrelenting geometric grid underlying all of it. Though this might allow an office to reach high population density, and may also help to minimize the distances between workers (which would seem to be an efficiency of a type), it will result in a space that is low in intelligibility. Not only will employees feel little sense of place in such an environment (new employees will become lost easily and may have unusual difficulty learning how the workflow in an office is organized if it is not signalled by the shapes of spaces), but desir-able traffic in ideas and information might be impeded as well.

The classic hive of cubicles is decreasing in popularity these days, as progressive companies work hard to find ways to maxi-mize retention of workers, especially in the knowledge industries that form an increasing part of the economy of the Western world. The basic cubicle design is still often a mainstay, though the manner in which its enclosing walls encourage or inhibit interactivity, and the effects of cubicle organization on workflow management, are garnering more attention than in previous times. Yet there is much work to be done to understand how space can be utilized to maximize productivity, economy, and job satisfaction. Some offices have tried moving to completely open designs in which employees are not provided with dedicated workspaces at all but are left to organize their own spaces using open tables and mobile technologies, perhaps with a few specialized walled areas to enhance privacy for smaller face-to-face meetings. Though such an open plan might work well for certain types of activities, especially for very small companies, it is less likely to be satisfactory for larger institutions, unless those institutions can rely heavily on mobile communications and are willing to encourage telecommuting. Both Cisco Systems and Hewlett-Packard have adopted such workspace plans; employees are encouraged to work from home (or Starbucks) whenever possible, and when their presence in the office is required, they simply put themselves in whichever part of the building requires their services. These companies report that they have realized efficiencies both in worker interactions and in the economic gain that comes of having smaller office footplates.[10]

Open, nonterritorial office plans are not necessarily a universal antidote to the cubicle design, however. One well-known example of the failure of such a design comes from Chiat-Day, an advertising company with offices in both Los Angeles and New York. In an attempt to increase collaboration, Chiat-Day removed dedicated workspaces and encouraged their workers to move around freely and to use different spaces according to their tasks.

Though workers did report increased communication, one of the goals of the new arrangement, they also complained of a lack of privacy, difficulty concentrating, and loss of time caused by the need to engage in searches to find particular people. Ultimately, Chiat-Day reverted to a more traditional design.[11] It could be that with better support from the mobile technology that has been developed over the past decade, nonterritorial office designs will be more effective, but it isn't clear yet whether the new technology has decreased our innate preference for face-to-face interactions.

In larger companies where such a free-flowing system of space use might not be possible, semi-open plans, in which the workforce is divided into smaller units, each of which occupies an open workspace, can produce satisfied workers with a strong sense of their place within an organization, provided that the spaces are well thought out. For one thing, spaces should be arranged to facilitate impromptu connections between members of unrelated work units. A common experience described by many employees of large companies is that the most innovative and exciting ideas can come about because of accidental meetings between people from work units whose functions may not be closely related. Using space syntax analyses, one can optimize a workspace to regulate the levels of such interactions. The use of space syntax to produce good social or thirdspaces, or even heavily trafficked corridors shared by multiple work units, can regulate the proportion of time workers spend in common areas where such valuable encounters might take place. Most social interaction does not take place in designed meeting areas such as coffee rooms or bullpens unless they are on well-integrated routes. As Judith Heerwagen and her colleagues put it in a review of the relationship between physical space and office work, "The pathway seems more important than the destination."[12]

Even simple proximity can have a major influence on our patterns of interaction within an office environment. One study of a large organization with two laboratories 60 kilometers from one another looked at the number of interactions between colleagues as a function of their locations. It was no surprise that almost all interactions were among people on the same floor. But what was surprising was that interactions between colleagues on different floors were no higher in frequency than interactions between colleagues in the two widely separated buildings.[13] One rule of thumb from early studies in the field suggests that those whose offices are separated by a distance of greater than 30 meters will almost never encounter one another spontaneously. Even this small zone of interactivity will shrink further still if the office environment contains many complex and unintelligible routes.[14]

One compelling example of how the organization of workspace can increase productivity and contentment comes from a study of the effects of an office redesign for ThoughtForm, a creative company involved in design, communication, and marketing that had relocated from one building to another in Pittsburgh. Figures 10 and 11 show the layout of the old and new offices, revealing some marked differences between them. In particular, the old office layout had a preponderance of isolated cubicles, which employees had sometimes noted as feeling "claustrophobic," and a notable absence of casual thirdspace in which those unplanned social interactions, particularly important to an organization whose product involved creative content, could occur. Indeed, the only spaces that were explicitly designed for group meetings appeared to be the formal conference and meeting rooms, both set well apart from the main working areas.

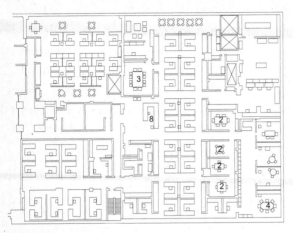

1. Main square
2. Project room
3. Conference room
4. Meeting room
5. Workshop
6. Library/relaxation room
7. Storage unit/hub
8. Reception

Figure 10: Original office layout

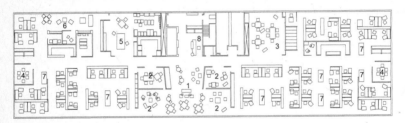

Figure 11: Redesigned office layout

In contrast, the new design featured a long central hallway or "main street" that increased not only the spatial legibility of the entire office but also the likelihood of unplanned hallway encounters. As well, the "main square," located in the center of the plan and directly opposite the reception area, was designed as an explicit thirdspace that could be used for anything from coffee breaks to PowerPoint presentations.[15]

Employees of ThoughtForm found that the new office design afforded enhanced opportunities both for privacy and for social interactions without any of the claustrophobic feelings of the former environment. (This is especially significant considering that the new office was 2,000 square feet smaller than the old one.)

Although it can be notoriously difficult to measure productivity in knowledge industries, especially those with a large creative component, there were clear signs that the new office design was enhancing the functionality of the company. Records of billable hours indicated that certain aspects of projects took less time once the company moved into their new quarters, suggesting that the new design, by enhancing social interactions and worker satisfaction, was increasing the company's productivity.

Judith Heerwagen has urged some restraint in the general trend to dropping the cubicle design completely in workplaces with open or semi-open designs. The challenge, she says, is to strike the right balance between the needs for collaboration and for quiet, private working spaces. Though many studies have found that benefits accrue from increased interactions, it is likely that the quality of individual work will suffer from the increased noise and distractions in the open environment. Heerwagen suggests the possibility of producing what she has called the "cognitive cocoon," which can surround workers with the tools they need to work without cutting them off from their surroundings.

How the balance between privacy and interaction is managed must also take into account the specific products that are being generated by an organization. Careful consideration must be made of the roles and the timing of individual work versus group interactions in a work process if a physical space is to be properly tailored to a company's needs.

For reasons that aren't well understood, companies that retro-fit space originally designed for other purposes often arrive at the most interesting and efficient workspace plans. One reason for this may be that the classic office tower, with standardized footplates on each floor, constrains thinking about how best to organize space and workflow. Retrofitting a space is more likely to require deep thinking about how to co-opt the size and shape of rooms devised for other purposes, and the outcome of such problem solving may be more likely to be a creative and satisfactory workspace design. One of the most beautiful examples of such a retrofit that I have seen is the University of Waterloo's School of Architecture, which moved into an old textile factory. Because the factory was built before indoor lighting was common, and because textile work demands attention to light and color, the building was designed with huge windows and skylights. The School of Architecture took advantage of these features, restored using modern materials, and adapted the wide corridors once used for moving large loads of fabrics to create a vibrant, exciting, and dynamic work and learning space with many effective thirdspaces.

Regardless of its size, location, or objectives, any work organization requires some kind of physical workspace, and the way that this space is arranged will affect the manner in which employees work, interact, and feel. Much of this influence of space on behavior follows from exactly the same principles that we saw applied to understanding how we behave inside our dwellings, and these principles in turn derive from the psychological nature of our connection with physical space. At heart, we are slightly odd creatures who collapse spaces into simple topologies, often telling ourselves stories or fitting ourselves into larger narratives in order to understand where we are. None of this might make much sense to an ant, a butterfly, or a honeybee, but it is a system that arises from the unique constitution of our brain, and it has consequences that range from where we go for comfort and refuge to how we earn our paychecks.

Whether they are single-family dwellings in the suburbs or gigantic architectural monuments in the core of a large city, buildings do not exist in isolation. They are collections of structures that produce the larger built environment of the street, the neighborhood, or the city. In some ways, the principles that determine our behavior in these larger domains are simply scaled-up versions of those we have seen operate inside buildings at the interface between constructed space and the fabric of our mind. In other ways, the larger canvas of the street and city produces an entirely new set of spatial concerns for us.

CHAPTER 9
CITY SPACE

How Knowing (or Not Knowing) Our Place
Influences Life in the City

*Space and light and order. Those are the things that men need
just as much as they need bread or a place to sleep.*

LE CORBUSIER

In the fall of 2005, Zyed Benna and Bouna Traore, two teenagers of North African descent, cowered in an electrical substation in Clichy-sous-Bois, a suburb of Paris, hiding from the police. The boys had done nothing worse than to engage in an impromptu soccer game with a few friends when they spotted a patrol car parked across the road from the playing field. Fearing that they would be detained, searched, asked to provide identity papers, and held, possibly for hours, at the police station, the boys fled. Benna and Traore, along with a friend, tried to squeeze themselves behind a big power transformer to avoid being spotted, but both boys, making fatal contact with unshielded wires, were electrocuted.

As the news spread of the deaths and the rumors of police persecution found wings, increasingly large numbers of young and dispossessed residents of the Parisian banlieues, oppressive suburbs filled with monotonous concrete-block buildings and largely occupied by the economically challenged ethnic minorities of France—mostly North African Muslims and Roma—took to the streets in protest. Over the succeeding three weeks, there were almost 3,000 arrests as rioters destroyed buildings and burned more than 8,000 vehicles, causing well over €200 million in damages.

The flames were fanned by reports that France's controversial hard-line minister of the interior, Nicolas Sarkozy, had suggested that the way to solve the problem was to remove all of the "foreigners" from these troubled streets. Given the economic hardship typical of first-generation immigrants, the alleged police discrimination, and the systematic persecution of minority groups by certain sectors of France's right-wing national government, it is easy to reconstruct a set of plausible causes for the widespread incendiary reaction to the deaths of the two boys. But one element that has received less attention is the built environment that was occupied by those who participated in the violence—that is, the ability of buildings or even neighborhoods to shape collective or individual human behavior.

At the time of the unrest, Clichy-sous-Bois was occupied by almost 30,000 people, among them some of the most impoverished in all of France. Not only was the area effectively isolated from the rest of Paris by an almost complete absence of public transport, but the streets were flanked by long, high, concrete buildings. Street intersections were rare, discouraging the flow of pedestrian traffic and minimizing any sense of privacy or ownership. The organization of the streets in the banlieues could not have been more different from those found in a traditional Muslim city like those in the countries of origin of most of the residents. Muslim urban centers, with their houses that face away from public thoroughfares and their graceful courtyard designs, emphasize privacy, family hierarchies, and clear lines of separation between public and private spaces.

In compelling images, Nico Oved's photographic exhibition "L'Habitat marginalisé" depicted the extent of the gloomy walls of concrete that surrounded the rioters and their families. Oved suggests that the shape and appearance of the streets themselves could have contributed to the conflagration in the streets.[1] These suburbs were a hangover of a damaging movement in architectural and city design that had been set in motion by Le Corbusier, the influential French designer and architect of the mid-twentieth century.[2]

In a misguided effort to find ways to house large numbers of people in small areas of space, Le Corbusier approached the problem by conceiving of the dwelling as a kind of a machine. Applying principles that eschewed ornamentation and adhered to mathematical principles based on the size and shape of the human body, Le Corbusier hatched designs consisting of vast plains of skyscrapers like honeycombs, filled with spartan but efficient living quarters and suspended on pillars to allow uninterrupted pastures of green space on the ground. Between the skyscrapers ran gigantic freeways that could move people rapidly and effortlessly from one part of the city to another. In the 1920s, when his ideas were incubating, the automobile was at the very beginning of its ascendancy, and in it Le Corbusier saw the solution to one of the key problems in urban development—the movement of people. In order for a large city to have a high level of dynamic integration, for all of its citizenry to have access to as many of the city's offerings as possible, there must be a way for large numbers of people to get from place to place in a hurry. The problem was, as urban visionary Jane Jacobs first pointed out, that to use the automobile to effect such movement,

Le Corbusier's numbers simply didn't add up.[3] His vast green pastures lying beneath skyscrapers would need to have been endless gray plains of parking lots. Le Corbusier proposed in all seriousness to the administrative officials in Paris that they raze large neighborhoods of Paris to erect his sky-kissing vision of the future. Fortunately for today's Parisians, Le Corbusier's grand vision was not taken seriously by those with the power to make such sweeping changes, but the influence of his ideas can be seen in smaller enclaves within Paris as well as in many other parts of the world.

HOW NOT TO BUILD A CITY

There are plenty of examples in our recent history of how neighborhoods fail when they are designed according to the ideals ascribed to Le Corbusier. Though his vision was not well realized in Paris, many of his principles were taken up in the design of large housing projects in North America. Indeed, because of the central place of the automobile in Le Corbusier's designs, and because most major urban development in the New World has taken place after the invention of the car, it has sometimes been too tempting to include elements of this kind of modernist, mechanistic design in our cities. Europe's large cities grew up over hundreds or even thousands of years, in a time when transportation of people and goods was either bipedal or, at best, by horse and cart, and so they have been more immune to the dramatic sculpting of urban space that the prospect of car travel makes feasible.

The reasons for the failure of modernist urban plans go far beyond an over-reliance on the automobile to solve problems of scale, though, and have much to do with the psychology of the urban dweller. Space in cities, much like the interior spaces of buildings, affects our behavior in two main ways. First, the organization and appearance of spaces can have a direct and measurable effect on how we feel. Second, the organization of city spaces can influence how we move, where we go, and how large numbers of people distribute themselves in the plexus of streets and thoroughfares in the urban mosaic. What makes all the difference in a city, compared to the smaller spaces of buildings that we considered in the last chapter, is that the people whose behavior is being influenced by the shape of the city are mostly strangers to one another. It is one thing to understand how the design of a dwelling can support hierarchies or gender relationships among kin, but another thing entirely to see how city design influences interactions among thousands or even millions of people who may cross paths every day or only once in a lifetime.

Nobody understood the importance of this difference better than Jane Jacobs, the urban visionary, activist, and writer who spent much of her life fighting modernist forces poised to reshape New York City and used her later years to exert similar powerful influences on Toronto, her adopted home. In her trailblazing book *The Death and Life of Great American Cities*—still current more than forty years after its initial publication—Jacobs offers a scathing indictment of the influence of modernist principles on urban design, but the more enduring contribution of this book is the collection of worldly wise prescriptions for designing livable, safe, and vibrant neighborhoods. Central to Jacobs's thesis was the oft-repeated mantra that "life attracts life." What cities need to stay alive is a multitude of public places filled with people. In part, this means that these public places, city sidewalks especially, must present opportunities for work, pleasure, and recreation, but it also means that the spatial organization of the city must accommodate such uses. Sidewalks must be wide enough that they can support pedestrian traffic as well as lingering. Blocks must be short enough to encourage pedestrians to make short hops from block to block. Collectively, these principles bring traffic, "eyes on the street" in Jacobs's words, but they do much more. By bringing together strangers with partially overlapping goals, they encourage a sense of shared ownership of public spaces. Because all see the value of the space, all defend its physical integrity and a code of behavior that maintains it. Because of this organic network of trust and understanding, the streets become almost as safe as the interior of the family home.

The importance of this sense of shared space was nowhere better illustrated than by some of its most striking failures. The infamous public housing projects in the United States, designed along modernist lines to provide low-cost subsidized housing to the needy, demonstrated how dire life could become in an area where poor arrangements of space served to break down social networks. Oscar Newman, architect and president of the Institute for Community Design Analysis until his death in 2004, described life in one of the worst of these developments, the Pruitt-Igoe project in St. Louis. Designed very much as a Le Corbusian plan, with skyscrapers suspended over verdant fields and trees, the development never reached more than 60 percent occupancy. The corridors and stairwells became deadly cesspools of human waste and garbage, and the flowing greenspace beneath the skyscrapers became "a sewer of glass and garbage."[4] About 10 years after its construction, the Pruitt-Igoe project was torn down.

In a lifetime spent trying to resurrect such projects and older neighborhoods on the slide toward slumhood, Newman identified what he perceived to be the key ingredient in the deficiency of such designs: the failure of residents to take ownership of public space. Newman's renovations included measures designed to place portions of public space into the hands of a few owners rather than many, believing that when a space is owned by all then it is perceived to be owned by none and therefore all rules are suspended. Like Jacobs, Newman understood the power of the eye. Spaces will remain secure only when there are many to watch over them. Jacobs believed the key to putting eyes on the street was to give those eyes reasons to want to be there. Newman, instead, tried to leverage the power of a sense of ownership by arranging space to encourage residents to feel a sense of pride, possession, and nurturance over the physical terrain just outside their thresholds. In both cases, though, these visionaries understood at a gut level what psychologists are now revealing about the importance of views and vistas as the foundation of the human understanding of the geometry of space. More important, they understood some of the ways in which our tendency to understand space as a set of connected views, rather than as a strictly geometric grid, could be used to encourage gathering, affiliation, and safety.

FEELING THE CITY

Those of us who live in cities (that is, almost all of us) might always have some background awareness of issues of safety and defense. Every street-smart person in a big city, even very livable ones like New York, Melbourne, or Toronto, knows that limits to one's freedom are imposed by standards of good sense. We don't linger in a deserted downtown alley at four in the morning unless we're hoping for trouble. But although we must think of our own safety in cities from time to time, the happy truth is that for most of us, most of our decisions about where and how we spend time are not governed by a concern for bodily safety. Simple preference also plays a large role in our movements and decisions.

Christian Nold, an artist by training, designs beautiful multimedia demonstrations of place preference in urban settings that he calls bio-maps. Nold sets up exhibits in urban centers in which volunteers are recruited to walk the streets of a city while wearing a small pack containing a suite of electronic hardware. The two main components of the pack are a global positioning device that can continually record the location of the wearer and a small machine that records a property of one's skin referred to as the galvanic skin response, or GSR. GSR is recorded by passing very low-intensity current between two of one's fingers and measuring how easily the current flows (though this sounds as though it might sting, the currents are so small that they are undetectable). A long tradition of experiments in psychology has shown that GSR readings correlate with arousal; this is exactly the same principle that is involved in using polygraphs as "lie detectors." In a way, we wear our hearts on our sleeves, or at least on our fingers, and GSR measurements reveal our feelings to the world.

Nold's participants walk around an urban neighborhood wearing these devices, soaking up the sights and wandering at will. When they return to him, the equipment they have worn provides Nold with a comprehensive record of their travels, their times of movement and lingering, and, most interesting of all, their emotional state at each step along the way. Though Christian Nold's bio-mapping initiative is designed more as a participatory performance piece than as a scientific endeavor, his results have much to say about human engagement with urban spaces. Overlaid on satellite photographs of streetscapes borrowed from Google Earth, the bio-maps show a cartographic sculpture of how the city feels. High arousal can be found at locations of stress (busy street crossings, for example) but also where the attention and interest of the walker has been engaged by a beautiful architectural facade, a busy market, or an interesting shop. Low arousal can be produced by large, empty spaces or oppressively boring facades. We all know that city travels produce these kinds of highs and lows in us, and perhaps even dictate our routes and stopping places, but quantifying these sometimes ephemeral states has been a difficult and rarely attempted task.[5] The much more common methodology among environmental psychologists has been to study preference for vistas of buildings, streetscapes, and landscapes by asking participants to rank photographs of them. Though these studies have generated interesting findings about our perceptions of the environment, there is much less evidence that our movements through cities and landscapes are correlated with our rankings of photographs.

Everybody who has lived in a city knows that each area has a distinctively different "feel" to it. Even when the purpose of our trip is specific, we may make choices that take us out of our way either to avoid areas that we find unpleasant or stressful (streets with too much traffic, quiet streets that we might find menacing) or to seek out areas of pleasure (busy pedestrian malls, attractive trails through urban parks). Even entire cities can seem to have their own personalities. I always see Toronto as a city of earnest virtue, like a little brother who tries just a touch too hard to act as if he were grown up. Vancouver is a shimmering, laissez-faire paradise by the sea with nothing to prove, its streets beckoning one to abandon serious worldly pursuits in favor of long runs through the park. New York's majestic skyline, with breathtaking architectural landmarks visible from almost every street corner and wide avenues filled with honking taxis and bustling pedestrians, strikes me as the lanky and attractive distant relative, impossibly smooth and sophisticated, yet with a strong, warm, and welcoming heart. Where do such impressions come from? It seems unimaginable that their origins could be captured by asking questions about pictures, recording the conductive properties of skin, or by sticking our heads into a brain scanner.

In his mammoth work *London: The Biography,* Peter Ackroyd struggles to capture a detailed impression of the grand old city by framing its history around its labyrinthine streets and alleyways. "Chapters of history resemble John Bunyan's little wicket gates," he says, promising to lead us "from the narrow path in search of those heights and depths of urban experience that know no history and are rarely susceptible to rational analysis." He makes good on his promise by taking us through almost 800 pages of a history entwined around space and place.

In a chapter on Fetter Lane, an avenue heading away from busy Fleet Street, Ackroyd documents the many uses of the space extending back to its fuzzy beginnings more than a thousand years ago. Though much of today's lane would be unrecognizable to a nineteenth-century resident, some common threads persist. The lane has always been a border. It was where the Great Fire of 1666 stopped. It was a mixed lane of brothels, taverns, and itinerant businessmen. It was filled with residents who lived on the edge, a good number of them tumbling over it when their crimes amounted to enough to warrant summary execution on gallows hastily erected mid-street. Though present-day Fetter Lane is filled with offices and sandwich shops, it retains enough of the old marks, twists, and turns that the ancient spirit of the street and its influence on our feelings may still pulse faintly, resisting the steady pendulum beat of the wrecker's ball.[6]

Ackroyd is a modern acolyte of an older tradition of psychogeographers, mostly philosophical and literary figures who felt, and tried to describe, a connection between feeling, space, and history. Like Ackroyd, psychogeographers share a sense that those hard-to-identify impressions that we glean from cities, neighborhoods, or even individual streets have much to do with the manner in which the shape and appearance of space have influenced their uses. Collectively, this amalgam of activity and geometry continues to resonate in the present time, influencing our feelings and actions as we walk a city.

The origins of the psychogeographic enterprise take us into the fascinating but bewildering territory of mid-twentieth-century French artistic and intellectual movements. One early proponent of the idea that city spaces evoke feelings as surely as mixtures of chemicals produce drug effects was Ivan Chtcheglov. In his "Formulary for a New Urbanism," Chtcheglov wrote that cities were inhabited by ghosts created by combinations of "shifting angles" and "receding perspectives" that "allow us to glimpse original conceptions of space." Central to Chtcheglov's methodology was the dérive, a kind of unstructured wandering where one was led from place to place like a robot being carried along the streets by simple rules related to the appearance of space. Though some of Chtcheglov's pronouncements were interesting, his contribution to our understanding of how we experience space was badly impeded by a steadily worsening mental derangement. Ultimately, Chtcheglov was institutionalized in part for plotting to blow up the Eiffel Tower because the light from it shone into his bedroom and disturbed his sleep.

Guy Debord, a clearer thinker than Chtcheglov, offered the first decent definition of psychogeography as "the study of the precise laws of and specific effects of the geographic environment … on the emotions and behaviour of individuals." Debord saw psychogeography as an essential element of a larger political initiative steeped in Marxist doctrine and intended to remake much more than the aesthetics of Parisian street corners. Debord's psychogeographic movement drifted into obscurity, in part because, as Merlin Coverley points out in his witty summary of the movement, the principal players spent far more of their time involved in definitional infighting than in actually carrying out the dérives that would be required to collect any data. Later in his life, Debord appears to have lost faith in the psychogeographic dérives to reveal any underlying principles relating space and feeling, preferring to believe that each person's relationship with urban space was intimate, personal, and beyond the reach of the conventional tools of science.[7]

IMAGES OF CITIES

Kevin Lynch, a leading twentieth-century American urban planner and a student of Frank Lloyd Wright, took a considerably more scientific approach to the general question of how cities are represented mentally. Lynch's concerns were more closely focused on the intelligibility of a city for the purpose of finding one's way around than they were on how cities make us feel, yet he saw close connections between the two. Lynch is best known for his groundbreaking work *The Image of the City*, the culmination of five years of survey, observation, and questioning of the residents of three major American cities, Boston, Jersey City, and Los Angeles.[8] Lynch defined the imageability of a form as that which made the form memorable. Highly imageable forms, by virtue of their "shape, color, or arrangement," were able to generate powerful mental images of these forms. Extensive questioning of interviewees led Lynch to propose that cities contained five main elements: paths, nodes, regions, boundaries, and landmarks. Paths are linear routes that people normally use to travel from place to place. Nodes are places of gathering, the places where things happen. Regions are sections of a city (SoHo, Beacon Hill, the Annex, Chinatown) that can be mentally represented as a single chunk of space. Boundaries are linear elements that are not used as paths, such as the edges of lakes or rivers, freeways, or railroad tracks. Finally, landmarks, as we have seen before, can be local elements that symbolize the meaning, use, or feeling of a part of a city or they can be large structures (such as the Sydney Opera House or Manhattan's Empire State Building) that are visible from many places and can act as navigational aids.

The main way in which cities varied, Lynch argued, was the extent to which each of these five elements were imageable, and he argued that the tools of urban planning could and should be used to influence this imageability. Chief among the evidence for imageability of city elements was the frequency with which they showed up on sketch maps of cities that Lynch asked people on the street to draw for him. Though many features in each of Lynch's three studied cities did seem to have high imageability, these sketch maps showed little sign of metric accuracy. The maps were topologically accurate, but distances were distorted, often in such a way as to suggest that the mental representations of the more imageable elements were expanded considerably.

Lynch's work takes us away from descriptions of how city space works that are concerned only with attraction, preference, and feeling and toward a consideration of how we cope with problems of space and wayfinding in cities. Although he emphasized that the strong impressions made by highly imageable city spaces are likely to be pleasant ones and well worth pursuing on entirely aesthetic grounds, another major part of his argument was that highly imageable cities are friendlier for the wayfinder. A navigator confronted with a morass of indistinguishable streets with little visual appeal or interest not only is likely to feel unhappy to be there but is also much more likely to become lost than a navigator in well-structured, highly integrated space filled with prominent visual features.

An entirely different approach to understanding urban behavior from the intensely introspective walks of the dérive or the analysis of sketch maps generated by residents is to adopt the simple observational methods used by many other behavioral scientists, including those whose work was described in earlier chapters on animal behavior. If we want to understand what people do and how they feel in cities, perhaps there's nothing better than to take up a position, as unobtrusively as possible, and to just watch them. This approach was pioneered in the United States by William Whyte.

After graduating from Princeton in 1939 and serving in the U.S. Marines during World War II, Whyte landed a good job at *Fortune* magazine and quickly rose through the ranks to an editorial position that gave him a chance to record in print his astute observations on the sociological upheavals occurring in postwar North America. From these writings eventually emerged the enormously successful book *The Organization Man*, whose recognition allowed Whyte to leave *Fortune* and devote himself full time to the study of urban behavior. In 1970, Whyte formed the Street Life Project, a small group devoted to using firsthand observational methods, including unobtrusive time-lapse photography and simple turnstile counts showing where, when, and how people spent time in cities.[9]

One of Whyte's first observations was deceptively simple. New York City contained a number of pedestrian plazas. Some of them were built at great expense, yet were always empty. Others were always teeming with business. Why? Whyte's group tackled the question with characteristic straightforwardness. They mounted high cameras to record movements of people, and they waded into the crowds to ask questions. Why had people come? Where had they come from? Why had they chosen this particular place? Like Jane Jacobs, his finest and most successful student, Whyte learned that life attracts life. In spite of what many builders had believed, people do not look for out-of-the-way, secluded spots in cities. More than anything else, we are fascinated by each other. We want to be as close to the fray as possible. Plazas that are close to major navigational routes are much more likely to be used than those that aren't. In a cycle of positive feedback, a small nucleus of people draws in a further influx of bodies as if by magnetic attraction. In no time at all, one plaza is filled while another, perhaps offering all the same amenities (except for the people), remains unused.

Whyte's observation, and many others like it, suggests that, just as we saw in the interior spaces of our houses, offices, and institutional buildings, the shape of space has a formative influence on where we go and how we gather. Route and enclosure move us from place to place with the same sureness with which a well-designed decanter pours oil onto a plate.

In his book *Architecture and the Urban Experience,* Raymond Curran uses exactly this analogy to explain how space moves us. Curran argues that all exterior spaces can be divided into receptacles, or "cluster spaces," in which we gather and avenues, or "linear spaces," that bring us to these gathering places. We are led along paths by our eyes, so artful presentation of contours, surface patterns, and colors can draw our attention with all the careful deliberation of a magician using sleight of hand. Spaces filled with horizontal contours sweep our eyes along, pulling our body behind them. High building facades and towers interrupt this sweeping gaze, causing us to slow and linger.[10]

Curran's illustrations make a convincing argument that in the design of cities, the spaces that lie between buildings are every bit as important as the buildings themselves. Danish architect Jan Gehl made similar arguments in his classic book *Life between Buildings* and then went on to show the city of Copenhagen how to put such principles into effect in spectacular fashion.[11] In the 1950s, like many other cities all across the world, Copenhagen found its dense core becoming choked with cars and empty of public spaces. Beginning with the placement of sidewalk cafés in the 1960s, Copenhagen's city fathers began a protracted campaign to resurrect public space in the city. With Gehl's help, this campaign reached its height with the opening of a major car-free pedestrian zone, the Strøget, which attracts throngs of visitors and residents and has become the crowning jewel of the city. The tactics used by Gehl to invite people back into city cores on foot have been so effective that now other cities, most recently New York, are trying to emulate them. At some point a new verb, "to copenhagenize," has entered the lexicon to describe the measures recommended by Gehl. In addition to an outright ban on automobiles on certain key thoroughfares, these measures include dedicated bicycle lanes, public squares, and restrictions on heights of buildings to enhance both the appearance and the climate of central pedestrian zones.

Copenhagen has become a model city for those who are trying to find ways to break down our car-centric way of life and to bring people back into the city core not just for commerce but for pleasure. Much of this evolution has been due to the steady influence of Gehl's approach, which combines insights into human psychology and sociology with sound architectural and urban-planning principles.

UNDERSTANDING THE GRAMMAR OF CITY SPACES

The kinds of on-the-street observations made by Jacobs, Whyte, Curran, Gehl, and many others have brought about workable principles that can effect positive changes in the shapes and arrangements of urban places. If such principles could be embedded in a more thoroughgoing theoretical analysis of lived space, one that made connections to the psychology of spatial cognition, then what has worked so successfully in some of the world's great places could perhaps be applied more systematically to our ailing cities. Bill Hillier, an architect and planner at the Bartlett School of Planning in London, has pioneered exactly such an approach to the understanding of city space.[12] Hillier has uncovered some powerful mathematical principles that can help to predict how our movements through space are partially determined by its shape. As surely as a thin-stemmed pitcher can be used to mete out minute quantities of rare truffle oil, a large piece of real estate can be carved up in such a way as to pull us through space in any desired manner.

Hillier begins by characterizing urban spaces as little more than strings of beads, where the beads are two-dimensional polygons of space, and strings are the paths from one district to another. From such patterns of nodes and paths, Hillier derives a grammar of space that he refers to as space syntax. The measures Hillier describes give us a way of describing relationships between small parts of a space—a single piece of road, for instance—and the entire context in which it is found. Whereas Whyte or Jacobs could stand on the street and describe with convincing and witty prose the nature of the street life that hums in lower Manhattan, Hillier's space syntax methods can give us a precise set of numbers describing the relationship between, say, Washington Square Park and the surrounding Greenwich Village streetscape. These numbers can be used to make solid predictions about where and how many people will be found in the park.

One of the most useful of such numbers is what Hillier calls the coefficient of integration. Each of the lines on a map of a built space can be given such a coefficient, and it represents, roughly speaking, the average number of turns that must be made to get from any one place on the map to any other. The easiest way to think about this is to imagine a neighborhood that you are familiar with, beginning with the street on which you live. Now, mentally travel from your street to the nearest convenience store, and count the number of turns that you would have to make along the way. If you repeat this exercise, but for a very large number of different destinations, then you can calculate the average number of turns required to get from your starting point to anywhere else in a region (or a whole city if you like). The higher this average number, the less integrated your starting point and the lower the coefficient of integration.

If you calculate integration coefficients for an entire region, the numbers can be represented nicely by using the heaviness of the lines on the map to represent the integration coefficient of each street. The heaviest lines have the highest coefficients, or highest integration, in other words. Shown in Figure 12 is such a map for my own neighborhood.[13]

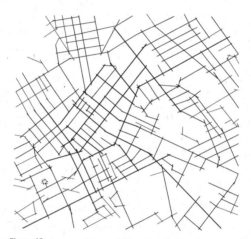

Figure 12: A map of my neighbourhood, showing areas of highest coefficiency using heaviest lines

The most interesting thing about these maps of integration coefficients, and the reason that I've burdened you with the details of how they are constructed, is that they work remarkably well to predict how we behave in space. People (and cars) tend to congregate at regions of high integration. In fact, such measures work so well that Bill Hillier's group has built a successful private consultancy that uses the principles of space syntax analysis to help cities plan buildings, streets, and neighborhoods in ways that promote desirable traffic patterns, both pedestrian and vehicular.

You might notice that there are some similarities between the analyses of space that I'm describing here and isovists. In both cases, some simple generalizations about the shape of space are being used to describe and predict how we behave in space. Notice as well that in both cases, we don't even need to describe *how* space is being used to make predictions about where we will find collections of people in a space. In the case of space syntax, there is a remarkable tendency for people to be found in the greatest numbers in areas of a city that have the highest integration values. The more connected an area is to the rest of the city, the busier it is likely to be.

Space syntax analyses such as these are not only useful tools for those in the business of designing city spaces. They also connect with our own mental maps of space. In one illustrative study, researchers knocked on doors in an area of London and asked respondents to draw sketch maps of their immediate neighborhood. At this point in the book, it shouldn't surprise you to know that these sketch maps were often wildly wrong and seldom represented distance and angle with anything like metric accuracy. Our minds simply do not map large-scale space in this way. What the findings did show was that when the sketch maps were subjected to the same space syntax analyses as the real places they were supposed to represent, there were remarkable affinities between the two. Though our mental maps may resemble real spaces only in a weak sense, they share the same syntax as physical space in the manner prescribed by Hillier's approach. Just as space syntax analysis collapses much of the metric size and shape of space to a series of nodes and lines, so does our mind put maps of space together as a series of simple viewpoints (the nodes) and the connections between them (the lines).[14]

Like isovist analysis, one of the most remarkable features of these studies is the finding that the reasons for our movements through a space are much less important than the bare structure of the space—the way that different areas in a space are enclosed and connected. Though knowing the functional organization of a space (where the stores are, where the washrooms are, and so on) can enhance our ability to predict movements through that space, the organization of the space is a much stronger predictor of our movements than what kinds of functions are served by the space. It might seem a little odd for me to tell you that when you jump in your car and drive into the city, or hop off the bus and start walking the streets of the urban core, I can predict exactly where you will go based on how the streets are connected together without needing to know that you have set out to, for instance, buy a pair of shoes. To make sense of this claim, remember that Hillier's theory is really meant to account for aggregate behavior of large numbers of people in cities. Though the computer that calculates the integration values of streets in your city couldn't possibly realize that you need new shoes, when many such trips to the city for many purposes are averaged together, the computer can do a pretty good job of predicting how such trips will be organized.

There's another way of thinking about the relationship between the shapes and connections of city spaces and the kinds of attractions that might be found in particular places. Imagine that you are an entrepreneur and you want to establish a retail business in a city. How do you look for a site? It doesn't take a roomful of psychologists or planners to tell you that you want to place your business in an area with lots of foot traffic, and it may be that the best way to find such a site is to use the time-honored methods of William Whyte—go and watch what happens on the street. In other words, it's probably no accident that successful businesses that depend on walk-in traffic are located in areas with high integration values. What is new and more interesting is that space syntax methods give us a way of predicting where such integration values will be found *before* we go looking, or even before the streets are built. These methods work so well because they mirror the ways that our mind responds to the properties of space, thus making a direct route from the way our mind works to the shapes of our cities. Our cities succeed or fail depending on how well we understand and manage this deep connection between mind and metropolis.

THE PSYCHOLOGY OF SPATIAL INTEGRATION

It may seem strange at first that space syntax takes little account of the distances between locations in a city. In our busy lives, we tend to find the shortest route from one place to another, especially as our familiarity with a city increases. Yet in Hillier's space syntax, it is only the number of changes in direction that determines the spatial integration of a location. Because Hillier's methods are designed to predict where people will congregate rather than to explain the wayfinding performance of individuals on single trips, the absence of distance measures in space syntax is not at all problematic in this respect. More surprising, though, is that in spite of our best efforts to plot efficient courses through the city, we are often fooled about distance in ways that are exactly consistent with the space syntax approach.

Many studies have shown that people who are led along walking or driving routes that have many changes of direction are likely to overestimate the distance that they have traveled.[15] Even though we are heavily invested in finding shortcuts, our spatial mind mirrors the nodes and lines of Hillier's spaces to such a degree that we confuse the dimensions of space in much the way that space syntax predicts. Just as we've seen on many occasions now, the rules of mental space often appear to be topological rather than metric, and so they are well reflected by descriptions of physical spaces based on topology.

Just as there is evidence that we have an intuition for the configuration of space given by isovists (one piece of evidence being our facility with finding locations of prospect and refuge in new spaces), there are also signs that we have a deep latent sense for areas of high integration in more complex spatial configurations.

Imagine that you've landed in an unknown city. You don't speak the language or understand the signs, and must rely only on your inner navigational senses to find your way from place to place. Over time, your movements are likely to spread slowly from a home base (perhaps your hotel) to other areas of interest—the place with good coffee, the best sidewalk bench for people watching, the nearest Internet café for catching up with the news. Often what happens is that you learn the configuration of a major street first, and then slowly learn to find your bearings from this major street to a growing set of goal locations. It is as if the major street becomes a kind of backbone upon which you build a more extensive skeleton of routes. What makes this strategy so effective is that the major street is likely to be a location of high integration, well connected and intelligible. As long as you can find your way back to this street from a variety of locations, your risk of becoming unfathomably lost is reduced. And because the major street is well integrated, your chances of finding it, even if you become disoriented, are high. Even if you lose your skeleton, the shape of space encourages you to rediscover it by wandering.

Notice, as well, that it is entirely possible to use this kind of approach to navigation without having any idea how the different goals you seek in the course of the day might be related to one another. You might have no clue how to walk directly from the Internet café to the coffee joint, but you know that the two places are connected via your skeleton, and that is all that you need to know to navigate from one to the other with ease. It isn't that the skeleton helps you to know the spatial relationship between the coffee joint and the café.

This relationship doesn't matter so long as you know that both places connect to the spine route. It's as if the café and the coffee joint are in two different universes, but they're connected somehow by the skeleton. There is a resemblance here to the findings I described in the first part of the book related to what I called the regionalization of space. We have great difficulty in drawing connections between the visible and the invisible in cities, just as we do in small sets of spaces in psychology laboratories. We learn useful tricks that can guide us from one space to another, often by using idiosyncratic links between places. There may not be anything spatially pretty or efficient about the routes we choose, but they work well most of the time.[16]

Many people have had such experiences, even in fairly familiar surroundings, where they might discover that a route they have been taking for years and years (a drive to the office, a walk through a shopping mall) is neither the shortest nor the most efficient. We sometimes even seem to take great pleasure in engaging in lengthy debates about route choices. When meeting friends for dinner in a restaurant, the debate about how best to share the bill is often rivaled only by the one about the best way to get home.

If there were deeply felt psychological principles of space at work in the organization of cities, then one prediction would be that we could discern patterns in urban streetscapes that reflected those principles. If city plans were laid out according to the whims and wiles of wandering travelers, then I imagine that this is exactly what would happen. We could look at an aerial photograph of a city and read off the inner workings of the minds of the beings who carved streets into the earth. As everyone knows, though, this is not how cities are built. Some cities, especially those of recent vintage in the New World (Washington, D.C., for example), were planned from scratch.

One of my favorite examples of a planned city is Canberra, the capital of Australia. Canberra's design was the result of a winning entry in an architectural contest. Walter Burley Griffin, an American architect, couldn't believe his luck when he was informed that his design, hurried together while on his honeymoon in 1911, had been selected and that he would have a chance to build a city from the ground up. As shown in Figure 12, the organization of the city is almost crystalline in impression, with a long central axis linking perfect sightlines of a war memorial and the parliament buildings. The street plan is filled with radial symmetry, and the center is cut through with a large artificial lake whose original plan called for tight geometric lines but was modified to provide some welcome relief to the mind by including flowing, organic shorelines.

Figure 13: The planned city of Canberra, easy to navigate but lacking in character

Canberra is a city of stark, geometric beauty with such perfect visual alignment that one's breath quickens in response to the sheer audacity of the design, but my own impression when I visited was that there was something sterile and slightly artificial about the space. I felt myself to be a visitor in a giant urban museum piece rather than in the living, breathing crucible of life that we normally expect to find in a city. My occasion for visiting was to deal with some minor administrative matters at an embassy, and I couldn't help feeling that the form of the city was perfectly in accord with my business there. I negotiated the wide, empty streets in my rental car with little difficulty, found the office I was looking for without delay, and, once my business was completed, I found few reasons to linger.

When cities build slowly, over thousands of years, some interesting commonalities emerge that show urban spaces as reflecting pools for the shape of the human mind. Hillier's group has noticed that most cities grow in similar ways, taking on a shape that they describe as a "deformed wheel" in which the central core of the city is connected to the periphery by a series of spokes with high integration values. This pattern is easily visible in London, Rome, and Tokyo and, with a bit of scrutiny, can be found in almost all city maps. Figures 13 and 14 show the deformed wheel organization of London and Tokyo, with shading indicating the depth of integration of individual streets.

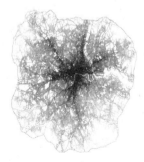

Figure 14: London's density forms the hub of a "deformed wheel"

Figure 15: Tokyo shows the same type of "deformed wheel" shape as London

Not only is this growth pattern the best way to promote continued contact between the central core and the periphery as the city increases in size, but it also makes large cities more intelligible to support good wayfinding. Hillier argues that this deformed wheel organization is driven in part by economic forces that favor minimizing distances between markets and buyers, and in part by the organization of our minds, especially the premium that is placed on viewpoints and vistas. Hillier contrasts his approach to understanding how cities grow with older methods based on concepts of mass and gravitational attraction. Such schemes suggest that places in cities are arranged in accordance with a system of attractive forces between individuals, groups, and institutions. Large institutions or social groups within a city space attract individuals as surely as the asteroid belt orbits the sun. In contrast, Hillier's space syntax approach is "light based rather than mass based . . . reflect[ing] the world we see rather than the world of distance and mass." In Hillier's scheme, it is not the unseen social forces of human networks of power that drive people through a space, but what attracts the eye.[17]

SPACE, CULTURE, AND TRAFFIC

Though most cities that have grown through gradual organic processes rather than through top-down planning show Hillier's spoke-and-wheel organization, it is obvious that there are enormous individual differences between cities, many of which are related to culture. One of the joys of world travel for me has always been the jarring confrontations with city plans that don't conform to the North American linear grid that I've grown up with.

On a recent visit to Beijing, I set out one morning to find a small museum. When I arrived at what I thought was the right address, I entered the building and began to look for exhibits. I encountered an elderly woman standing before an open fire stirring a pot. I tried to speak to her in Mandarin sounded out from a phrase book, but to no avail. My mouth became dry and I began to sweat. Without the ability to communicate verbally, I looked around for spatial clues as to the nature of the room that I stood in and the building that enclosed it.

The relationships of open and closed spaces resembled little that I had seen before. I had no spatial cultural reference. It took considerable effort for me to understand that I had stumbled into a private residence and not a museum. This is a small example of a general truth. When we embed ourselves in foreign cultures, the strangeness of how built space is used is just as disorienting as an inability to understand the local language. In Beijing, there were times when it was difficult for me to distinguish residence from business, or even residence from sidewalk—some dwellings appeared to consist of nothing more than a cloth awning stretched out over a sidewalk or even a roadway, with a cooking pot over an open fire constituting a kitchen, and some rolled blankets arranged around the outside of the canopy serving as bedrooms. Variations in spatial culture are not always so extreme, but they are persistent and often easy to spot.

Hillier has used the city of Nicosia in Cyprus as an example of the influence of culture on space. Like many cities, Nicosia contains separate ethnic enclaves. The Greek settlement has a rough linear grid of streets with high integration that would not look out of place in North America. The Turkish settlement, with low integration, short streets, and low intelligibility, is typical of many Muslim town plans.[18] Such plans are meant to steer visitors toward a few well-specified public areas and away from residential areas, and generally to encourage privacy and to discourage co-presence. These differences show that there is an interaction between our cognition of space, the manner in which the arrangement of space sculpts our movements and our behavior, and the requirements of a particular culture. *All* human beings are affected by the organization of space in similar ways, and we can use this generalization about our minds to organize our movements in ways coherent with cultural values. Skilled architects and designers can bring people together or keep them apart with the same precision that a skilled potter employs to make a jug designed to mete out single drops of precious oil.

———————

As cities grow, one of the main constraints acting on their form is the ease with which residents can get from one place to another. The deformed wheel suggests that one way to promote such ease is through artful design of the meshwork of streets, but in modern times this can take us only so far. As urban populations have increased to staggering numbers, we have found it necessary to seek ways to conquer space by cheating time. We have learned to move more quickly.

There can be no doubt that the advent of the internal combustion engine has had more impact on the shapes of our cities than any other single development in the last thousand years. The rules of spatial cognition don't change—indeed, space syntax studies have shown that we can predict the movements of cars with the same precision that we can foresee the movements of people. The same tools that help us to improve the Tate Gallery or fill a public square with pedestrians can also predict how cars will move in urban street plans. But cars change the scale of cities dramatically. One rule of thumb among urban planners is that people will walk when their goals are located less than a five-minute stroll from their houses. Though the exact number or distance (sometimes referred to as a walkshed) may be a subject of debate, and may vary depending on the demographics and motivations of the walkers, one thing is abundantly clear: a driveshed is much larger than a walkshed. This is not only because cars move much more quickly than pedestrians but also because driving is almost effortless compared with walking. Provided that it is easy for us to climb into a car, and provided that the layout of streets makes it practicable for us to get around at a reasonable speed, we are willing to tolerate vastly exaggerated distances between our homes, our places of employment, and the locations of services and stores that contribute to our happy lives. Although rapid transport using both private cars and public transportation systems has made it possible to accommodate huge numbers of people in urban settings (worldwide, there are more than 300 cities with populations exceeding one million people and a handful of cities with populations in excess of 10 million), it has transformed the nature of the city and given rise to a plethora of difficult problems, many seemingly without any reasonable solution.

From the spatial perspective, rapid transit by whatever means requires that cities be able to work at simultaneous but wildly different scales. While cars hurtle along expressways and wide multilane thoroughfares and trains filled with commuting workers plow through subterranean tunnels, many of the pedestrians whose needs have not changed much in a millennium are still wandering the sidewalks at street level. Accommodating the needs of travelers with such different spatial scales, and providing a streetscape that works for all, has been one of the most vexing problems ever confronted by those who try to plan cities.

Downtown areas work best when there is high density of use by pedestrians, yet if the pedestrians have no means of getting around in the city, they obviously cannot contribute to density of use. Though rapid transit should be one solution to this problem, it is not a panacea. For one thing, most types of rapid transit systems are enormously expensive and must be paid for from the public coffers. Furthermore, it can be very difficult to coax people out of their cars and into buses and trains. Many of us climb into a car for a daily drive to work even though using available public transit would be less expensive for us and, in many cases, require less effort. Many people end up parking their cars in lots that are much farther from their workplaces than the closest public transit stop.

One of the most remarkable demonstrations of how difficult it can be to convince human beings of the laws of geometry and physics has come from the major battles that have been waged over congestion pricing for urban roadways. In congestion pricing schemes, drivers who enter certain parts of a traffic-choked city during peak hours are required to pay a stiff surcharge. London, England, so far the largest city to have adopted congestion pricing, charges £8 to vehicles entering the city during peak times. When it was introduced, the scheme was enormously controversial, with central businesses claiming that they would suffer devastating losses and commuters arguing that in spite of a comprehensive mass transit system, they would have great difficulty getting to work on time. In practice, the effects of congestion pricing in London seem to have been milder and mostly positive. Though some central businesses have experienced significant sales losses, the decrease in the number of cars that visit central London on any given day has been much greater than the decrease in the total number of visitors to the city, suggesting that many more people are opting to use mass transit systems. There is also no denying the health benefits of the reduction in cars in the city. Levels of carbon dioxide and sulfur dioxide, two emissions products of car exhaust, are measurably lower.[19]

Mental distortions of space and time may also explain some of these strangely irrational preferences. Driving a car is more active than sitting on a bus, so time can seem to pass more quickly, hence contracting perceived distances. People have many other reasons for preferring private car travel over public transit. In New York City, Mayor Michael Bloomberg has been an ardent advocate of congestion pricing, but has had to face enormous resistance to an idea that not only makes sense but is supported by the London experience. Results from an extensive telephone survey conducted on behalf of the Partnership for New York City, a group that supports congestion pricing, showed that many people chose not to use mass transit because they believed that it would increase their travel time (which is almost certainly not true), they preferred to be in control of their own movements, and they wished to avoid contact with other people. There is some irony here: the same feature that draws people into public spaces (the desire to be near and to observe others) seems to actually repel them from mass transit systems. The reason may be the type of space involved in each case—the interior of a bus or subway car is very different from a bench in Rockefeller Square.[20]

In our own cars, we travel while enclosed in a small portable space that we feel we own. A car provides a sense of continuity from the private spaces of home all the way to the spatial threshold of the workplace. On a more pragmatic level, the car gives us more flexible mobility—if we choose to make a detour on our way home, this is much easier to manage in a car than when using public transport.

Savvy urban planners know that in spite of our intrinsic bias toward private car travel, there are many ways to encourage people to forgo their cars in favor of other means of transport, all of which rely in some way on bringing the spatiotemporal advantages and disadvantages of car transport and public transport into closer alignment. Narrower roads, limited access to freeways, and even the timing of traffic lights can tweak the times taken to travel between common destinations in cities, and there is much evidence to show that these measures work to shift the balance of how people move about in cities. Indeed, the early battles of Jane Jacobs that did so much to establish her reputation as one of the few voices of reason in a wilderness of howling insanity among those who built cities had much to do with such arguments. When cars are diverted from prized pedestrian areas within a city, they do not simply show up in other areas like leaking fluids seeking the quickest route to lower ground. When life is made more difficult for cars in an area of the city, the cars simply disappear. More people opt to ride buses and trains or to walk.[21]

Why is there a need to control traffic in cities? The main reason is that if there were no impediments to car travel, and public transport were not readily available, then our urban centers would become choked with traffic and the noise and pollution that accompany it. In addition, the numbers of cars traveling the streets would far exceed the available parking spaces in the city. Especially in North America, cities have tended to sprawl outward, thus exacerbating considerably the problem of coping with city scale. Because, in terms of total population in relation to geographic area, cities tend to be larger than they need to be, population density on the ground is thin. This low density contributes to the difficulty of providing proper public transit and makes it difficult for residents of a sprawling city to meet the ordinary needs of their lives without using private cars. Though this trend toward sprawl in cities has accelerated tremendously in the latter half of the twentieth century, it has its roots in the very beginnings of colonization of the North American continent by Europeans in the seventeenth century.

THE PSYCHOLOGY OF SPRAWL

In his seminal account of the origins of suburban sprawl, *The Geography of Nowhere,* Howard Kunstler points out that from its inception, the concept of property ownership in North America departed from some important European traditions. In Europe, ownership of land was deemed to be a public trust. A part of the trust was an understanding that one would undertake proper stewardship of the land for the common good. In America, on the other hand, land was considered much more strictly an economic resource. The main point of land ownership was that it could be leveraged into financial success. Because of this tendency to see land as a form of currency, much less attention was paid to the relationship between the topography of an area and its conceptual value in economic units. So land was subdivided in straightforward grids of regular geometric units, a pattern easily seen in many cities (New York is the classic example, with its regular and monotonous layout of city blocks). This manner of land subdivision, as we now know, not only ignores the importance of geographic features but also neglects the important contributions of our mental topography to the ways that we view, understand, and use city spaces.[22]

This way of thinking about land also nurtured fierce defense of individual jurisdiction over land use. If land is money, then nobody would want to be told how they should use their currency. These two factors—jealously guarded rights over land and a tendency to parcel it out without regard for the hills, dales, rivers, and streams of real geography and their influence on the human psyche and how we deal with space—conspired to produce rapidly industrialized cities, so that real estate could generate as much income as possible. As a consequence, urban areas became horrible places to live. Streets were packed with factories spewing all manner of noise, heat, and toxic wastes. Workers were packed cheek by jowl into tenements owned by slumlords who felt they possessed a divine right to treat their tenants to any amount of physical deprivation.

In light of the degeneration of the urban environment, those who could afford to do so looked for ways to flee from the city.

The first suburbs in North America took advantage of the advent of railroad lines that made it possible for the affluent to make the daily commute to a city workplace, but in many other respects these suburbs resembled some of those we might find in modern North American cities. Building lots were huge by city standards (an acre or more) and roads built for horse and cart were wide and winding; both features provided privacy for residents and helped to complete the illusion that they were living in the country. From the beginning such suburbs were designed to be free of mixed use. Any goods required by the wealthy suburbanites could be delivered to their door by courier. Meeting halls, gathering places, markets, or any form of public space was entirely absent.

The advent of the automobile changed much about the suburbs. For one thing, affordable transit for the masses served to democratize these areas by bringing them within easy reach of the middle class. Because of this, suburban areas have grown to enormous proportions around just about all major and midsized North American cities. Though much has changed in the way we think about life in cities, the main features of suburbs have not changed at all. They are replete with winding roads that enhance privacy but discourage pedestrians. The byzantine layout of streets, often named after the trees, plants, and animals that were uprooted to build them, is unintelligible in Hillier's sense. A more critical problem is that there is little to do in monofunctional areas devoid of public places. Houses in suburbs are designed explicitly to facilitate an automobile-centered style of life, so the streets are empty of pedestrians. The most prominent part of the house facade is usually the garage door, complete with a remote-controlled power door opener so that commuting homeowners can drive directly from the office parking lot to the interior of their living space without once making contact with the outside world. Those who dare to venture onto the streets on foot often face some peril, simply because pedestrians are so rare (not to mention difficult to see as drivers negotiate one sinuous turn after another) that they risk being run over.

Some years ago, my wife and I, having tired of running up and down the creaky, narrow stairs of our old urban house with many armfuls of children, decided to give life a try in an "upscale executive suburban home" as the realtors called it. The living was easy, comfortable, quiet, and ultimately soul destroying for us. The turning point came at a neighborhood meeting that was called to address the increasing traffic density in our neighborhood caused by drivers speeding past our houses to shave a few seconds off their commute to an adjoining new subdivision. The problem became so severe that we had to ban our children from playing in front of our house on the corner for fear that an out-of-control car would mount the curb and kill someone. When we suggested that the situation could be solved easily and cheaply by closing the crucial shortcut street at no detriment to any of the residents of the affected areas, a city official told us that such a solution would split the neighborhood. Residents on each side of the closed road would no longer be able to visit each other. When we suggested that those so affected could simply walk to one another's houses, the traffic engineer curtly replied, "They won't." The issue was closed. We put our house on the market a few days later.

In the second half of the twentieth century, urban sprawl accelerated dramatically. The urban area of Atlanta ballooned to over 160 kilometers when measured from north to south. Detroit's population actually decreased by 2 percent while its land area increased by 45 percent. And urban sprawl is not just a North American phenomenon. Brussells, Frankfurt, Munich, and Zurich all showed decreases in population density in the late twentieth century. Even

Copenhagen, a showpiece city for improving pedestrian access to the urban core, had a net decrease in population density.[23]

There is much to enjoy in the suburbs. Houses and building lots are often gigantic. Quiet cul-de-sacs can serve as safe play areas for children or even, if conditions are ideal, gathering places for adults. Provided one has access to a car, it can be easy to manage the great distances involved in meeting the day-to-day needs of life, such as finding groceries, health care, recreation, and entertainment. Wayfinding can present some challenges, but as such challenges are usually confronted while behind the wheel of a car, little real effort is expended when one becomes lost. Because of the lack of public spaces, making social contacts in suburban settings can be difficult. As one is less likely to make chance encounters with neighbors on the street, building local social networks can involve knocking on doors and making explicit invitations to share time in the private spaces of homes. Only the most gregarious individuals in such neighborhoods are interested or able to cross such thresholds.

There are many significant reasons for us to be concerned about the sustainability of this type of suburban development. Sprawl consumes vast amounts of agricultural land on the edges of cities, and disturbs watersheds, disrupting both ecologies and supplies of potable water. The horizontal expansion of cities encourages the use of cars. Apart from the links between carbon dioxide emissions and climate change, smog has deleterious effects on health. Smog-related deaths are on the rise in many major cities, and some estimates suggest that smog kills more people annually than do car accidents. Many believe that the time of cheap oil is coming to an end. Globally, we are beginning to demand oil more rapidly than it can be supplied. Whether or not one agrees with the doomsayers who suggest that we are at or just beyond the crest of oil production and are now poised for a downward slide into indescribable economic and social collapse, there can be no denying that fossil fuels will eventually run out. In the long run, a way of life that is built on the assumption that the supply of cheap fuel is endless is not sustainable.[24]

SMART GROWTH?

In response to this broad constellation of factors—a degenerating environment, research showing that our failing air is not only killing the plants and animals that we share a home with but killing us as well, and heightened awareness that the economic meaning of physical distance is strongly tied to the cost of the fuel required to get from one place to another—many regions are considering or have already adopted so-called smart growth agendas that call for limits to sprawl, urban intensification, enticements for mixed use in urban cores, and measures meant to lure people out of their cars and onto the streets. Can what we know of human spatial cognition and wayfinding contribute to such agendas?

A pioneering legislative scheme for decreasing sprawl was the State of Oregon's imposition of rules for urban growth boundaries in 1973. This incredibly forward-thinking act has had the effect of combating sprawl in, for example, the city of Portland, where population density has increased, the downtown core has remained vibrant and interesting, and watersheds and agricultural lands surrounding the city have been preserved.[25] This is not to suggest that there's paradise in Portland, however. In spite of the regulations, Portland's urban boundary has slowly stretched outward and there is still traffic congestion. In addition, Portland, like other cities that have encouraged urban intensification such as Vancouver and Melbourne, has seen exponential growth in property valuations.

A Canadian scheme for decreasing sprawl is the much-lauded Places to Grow Act that was passed into law in 2005 and has been awarded prizes by planning institutes in both Canada and the United States for its vision and leadership.[26] Among other things, the act sets density targets for combinations of jobs and residents in designated urban regions within the most densely populated part of Ontario. The act leaves it largely to the affected municipalities, along with developers, to work out how the density targets will be achieved. This approach, not an uncommon one in many similar acts in North America, has some risks. Past experience suggests that, other things being equal, people have a strong preference for expansive space. Average house sizes have ballooned from about 900 square feet in 1950 to more than 2,400 square feet in the early part of this century. At the same time, family size has decreased. In a way, this is a self-perpetuating trend. The spatial design of the suburbs discourages pedestrian activity and shared public spaces, so, for the resident of the suburbs, the interior private space of the home and the manicured fenced-in yard can become the entire world.

If intensified urban spaces are to be successful and not just enforced sardine cans full of human beings, like failed housing projects in the United States in past decades, then these intensified areas must be designed to encourage their residents to expand their spatial range. People must be encouraged to walk, not only by providing them with attractive nearby destinations such as pedestrian malls and public gathering places but by the very shape of the streets. A good urban vista can pull people out of their residences and into shared space. Street plans with legible space syntax can enhance wayfinding, promote rapid and accurate mental mapping, and so contribute to the pleasure and ease of the wandering crowds. Judicious planning of visual contours using building facades, and the manipulation of visible street and sidewalk width using trees or other adornments, can lead the eye in desired directions. The eye being lured, the feet will follow.

The dimensions of public spaces and thoroughfares can influence where we go. Likewise, the sizes and shapes of spaces can influence how we feel while we are walking. Not enough is known of the science underlying the experiences of the early psychogeographers or Christian Nold's modern wired equivalents to understand what properties of space contribute to the emotional experiences of the street, but some basic principles are clear enough. Few very large public spaces are successful, perhaps because they are all prospect without refuge. Large setbacks between facades and sidewalks do not work well. Not only does such an arrangement make the pedestrian feel exposed but the potential for rich visual detail as one walks past facades, shop windows, and other resting places is wasted. Buildings that are too high, closing in streets like narrow canyons, have a negative effect on pedestrian experience. The best street vistas contain some enclosure, street width and building heights conform nicely to human proportions, and interesting landmarks help to enclose views of the ends of streets.

Though much could be done to improve the human condition by rearranging bricks and mortar in city spaces, the simple truth is that our fantastically versatile brains have given us entirely new spatial vistas to play with. Most urban dwellers are likely to spend at least a part of each day sitting in front of glowing screens that provide them with portals into spaces that could not have been imagined when the grids of cities in North America, Europe, or Asia were being laid down. Our ability to design machines that can transport us instantly from one place to another using nothing more than beams of electrons has further revolutionized the human relationship with physical space.

CHAPTER 10
CYBERSPACE

How the Nature of Our Mind Makes It Possible
for Us to Live in Electronic Places

*People on planet earth are like a bunch of really
technically bright teenagers without any supervision hanging
out all summer in a chemistry lab.*

JARON LANIER

There's something about my face today that doesn't seem quite right. It may be a little too thin, or it could be that I've trimmed my beard too closely. My white T-shirt is form fitting—an unusual choice for me as I usually prefer to wear looser clothing to hide some of the extra pounds. Walking down a wide thoroughfare, I notice a large green box squatting low in the grass beside the road. It looks like one of those giant utility boxes found in some suburbs, but it is emitting a strange whooshing noise. For some reason, I decide to sit on it.

I don't think there is any risk here that I can come to much harm. I notice a woman wearing a long white flowing dress. She is sitting on a bench nearby. I wander over, wanting to engage her in conversation but wary of too close an approach. I don't want to frighten her or have her misunderstand my intentions. I know from some previous visits to this part of town that there are all types of characters around, not all with noble intentions. I ask her if she minds if I sit down and she mumbles a word or two of assent. The bench was a little farther from me than it looked, so I fumble awkwardly as I sit. When I try to engage her in some small talk she tells me that her English is poor. I hadn't noticed that she was Japanese. Two young men walk past us, staring in our direction. One of them mentions, with something like a chuckle, that we look as though we are waiting for a bus. He wonders whether to tell us that no such conveyance will be arriving. Some distance down the road, the two men pause and then come back to address the woman, asking her if she would like to accompany them to "go have some fun." They have bad intention written all over them, so I'm surprised when she gets up from the bench and wanders away with them, laughing. I wonder whether she has understood them. I'm even more surprised by a little pang of what feels like rejection and jealousy as I realize she's passed up a chance for a friendly chat with me on a peaceful bench in favor of an invitation to cavort with these bad boys. I watch the trio shrink in size and fade into the distance, and then I take flight, zooming straight upward to about 200 meters so that I can watch them from above. I don't think they notice me as they run from place to place in wild zigzags. There's not much else going on in this sector, so I teleport home and log off. It's time for bed.

Ten years ago, the story of my evening in an alternative universe where I could walk, sit, and talk but also fly and teleport myself to new locations instantaneously would have been taken as science fiction, but many readers of this story will recognize that I was describing something that can now be experienced by anyone with a connection to the Internet. Second Life, a commercial venture spearheaded by the company Linden Labs, was begun in 1999 by Philip Rosedale, a San Francisco entrepreneur who had grown up with a fascination for computers, electronics, and virtual reality. In technological terms, Second Life consists of a large bank of servers that are used to house simulations of a huge tract of virtual space. Individual users can obtain free accounts and, by installing some client software on their own computers, they can visit many of the areas of the large matrix of spaces offered up by Second Life. Though multiplayer games (sometimes called MMORPGs, for massively multiplayer online role-playing games) have been around for some time, Second Life is different. Its directors insist that Second Life is not a game but in almost every imaginable respect is as real as physical space. Second Life's many millions of users participate in a genuine economy in which millions of Linden dollars change hands every month. Linden dollars can be exchanged for real "out of world" hard currency. What do the residents of Second Life do? Pretty much anything they want. I've seen residents chat, dance, and gamble, shop, prance about on nude beaches, fondle, and fornicate. Serious users of Second Life hold seminars, business meetings, classes, and open houses. Large companies such as IBM, Intel, Dell, Microsoft, and Toyota maintain business offices in Second Life. The government of Sweden maintains a virtual Second Life embassy. The news agency Reuters, having successfully managed the transition from propagating information via homing pigeon to sending the news through fiber-optic cables, also disseminates Second Life news stories from a virtual office.

Rosedale attributes much of the phenomenal success of Second Life to a key decision undertaken in 2003, when prospects for the fledgling company looked bleak and layoffs of an already small workforce had begun. In keeping with the underlying philosophy that Second Life should be made to simulate a real world with a real economy, people were not only allowed to purchase and own virtual land but they were able to build, design, and invent virtual structures and machines in Second Life that could be offered for sale with all of the protections in place offered to those who vend goods in a bricks-and-mortar building in a physical marketplace.[1] What is revolutionary about Second Life is that, compared with popular games held in virtual spaces such as World of Warcraft, the company that offers the grid of spaces is not at all involved in the production of content. The entire digital planet, or metaverse as it is sometimes called, is designed and built by users themselves, sometimes with the assistance of entrepreneurial "in-world" professional designers who create entire complexes of buildings for payment in Linden dollars.

The universe offered up by Second Life might seem like the normal digital content that is provided by the Internet but in a new and spiffy guise. Rather than cascading through pages of Web content with fancy graphics, flashy animations, and little talking heads, Second Life offers the same kind of content but from inside a metaphor that pretends to mimic a few more features of real life.

Users are able to design their own avatars, collections of animated pixels that can be tailor-made to suit an individual's taste. Body size and shape, hair color, facial features, and clothing can be changed easily. Normally, a user's view of the virtual world is centered on a position just behind their avatar, so it is possible to observe one's own actions from a position slightly above and behind the center of one's locus of embodiment, but it is easy to manipulate the "camera" for a different point of view. Simple keyboard and mouse commands control movement, and avatars can communicate with one another either by typing on a keyboard so that text messages appear at the bottom of the screen or in some cases directly by voice.

It's a little too soon to assess the extent to which users of Second Life and other similar realms feel themselves to be immersed in a completely new metaphysic outside of reality, but there are some early signs that the virtual spaces of Second Life are more psychologically compelling than would be expected from a jazzed-up Web browser. Nick Yee, in work carried out for his Ph.D. thesis at Stanford University, conducted a novel study of the influence of what psychologists call proxemics. Edward Hall, an anthropologist, documented in the 1960s the manner in which we regulate the spaces between ourselves and other people as we go about the tasks of everyday life, including different kinds of social interactions.[2] Hall distinguished between personal and social distances, for example, meaning by the former the interpersonal distances enjoyed by close friends as opposed to those physical spaces we maintain between ourselves and our casual acquaintances. Since Hall's early work, many psychologists have studied the influence of other variables, including gender, on interpersonal distance. Yee's approach was to carry out the same kinds of measurements between conversing pairs of avatars in Second Life. He discovered that interpersonal distances were influenced by gender in much the same way as happens when individuals converse in real life. Men stand farther apart and engage in less eye contact than women when they talk, even in virtual space. (Although the simple graphics employed in Second Life rule out genuine eye contact, it is possible to manipulate your orientation so that you are either standing squarely facing another person or "glancing" off to the side.) When men do find themselves standing close together in a way that violates the rules of proxemics, they compensate by decreasing the frequency of eye contact even further. The importance of this finding is that it suggests that even in a virtual space, where there is no possibility we could ever mistakenly believe ourselves to reside in that space, we comport our digital selves in a manner befitting a belief that we really embody our avatars.[3]

Findings such as these will probably come as no surprise to those legions of fans of role-playing games such as Guild Wars or World of Warcraft. Though some of us reminisce about the early days of computer gaming, in which text-based adventure games and simple arcade games like Pong and Space Invaders were able to hold our attention until the wee hours, there can be little doubt that modern games with their elaborate scenery, beautifully rendered graphics, and engaging animated characters generate a genuine feeling of personal embodiment within the play of images on the computer's screen. Indeed, many have expressed serious concern about the addictive potential of computer games, and many gamers have reported disruptions to family and work life produced by an inability to turn away from the computer or the gaming console, or even a failure to properly distinguish between their real and virtual existences.

Much of the immersive nature of modern and sophisticated games has as much to do with the engaging narrative structure of the game as it does with the visual effects, but there is little doubt that the careful rendering of an artificial version of real space, complete with complex and believable scenery and objects that behave in ways similar to real life, also contributes to the game-playing experience. In some games, such as Guild Wars, the natural scenery, including mountains, forests, and water features, is so realistic that it would not be surprising if immersion in such virtual settings produced some of the same pleasant effects as a walk in the woods. In one study completed in my own laboratory, we found that exposure to virtual environments simulating natural settings produced convincing decreases in physiological measurements of stress using GSR (the same measure employed by Christian Nold in his biomapping initiative).

CONNECTING PLACES WITH ELECTRONS

Virtual spaces constructed using clever Web-based metaverses like Second Life are only the most recent progeny of a trend in the human transformation of space that began with Alexander Graham Bell's scratchy command to his assistant, Thomas Watson, to "come here," accelerated with Guglielmo Marconi's remarkable question sent in Morse code across the Bristol Channel, "Are you ready?" and began to take over entire realms of human consciousness when, in 1929, Pem Farnsworth appeared as a shaky image on the first television set designed and built by her husband, Philo.

Before any of these developments, the speed of communication between one human being and another had a hard upper, biological limit—the speed of a running human being, a flying bird, or a fast horse. With our conquest of the electron has come light-speed transmission of signals, beginning with sparse codes and scratchy voices, but now filled with images and interactivity.

Many warehouses could be filled with all the books that have been written about the far-reaching implications of these changes in how we send information from one place to another, and justifiably so. The instant, widespread dissemination of scenes that would normally be beyond our immediate visual grasp has completely transformed how we know about the world, and it has had an especially dramatic impact on human engagement with physical space.

In his influential book on the effect of new media on human experience, *No Sense of Place*, sociologist Joshua Meyrowitz uses an architectural metaphor to help readers begin to get a sense of how visual media such as television have influenced our lives.[4] Imagine, he says, that all the walls of our buildings simply vanished. There would be no more private or public spaces or any other influence of physical place on our social lives, nor on our perceptions of how the world is put together. We would no longer have the option of framing social interactions by putting them into particular physical locations. Meetings "behind closed doors" would no longer be an option. The advent of television has had a similar effect on our lives. Though the connections are one-way, any two places connected by a camera and a television set are linked as if by wormholes through space. Furthermore, the fact that television broadcast signals are propagated through the air (at least for a little while longer) means that the manner in which space is folded in upon itself by these invisible waves is completely democratic. Anyone who has the appropriate technology to receive the signals and who sets up equipment within range can obtain and view the content of the images. Without making a judgment about whether television has exerted a net positive or negative influence on our modern way of life, there can be no question that the steady flow of content through the airwaves has transformed us in almost every way. Most important, television has worked hand in glove with other types of technology, such as rapid transit, that allow us to move our bodies quickly through space.

Collectively, these human inventions have made huge inroads on the natural relationships between place, movement, and position. Now, I can transport myself from my living room to any other location on the planet for a live view of a war, catastrophe, concert, or sporting event. Benefit concerts, journalistic coverage of war zones by "embedded" journalists, World Cup soccer matches, and award shows bring unfathomably large numbers of viewers together into one single shared view of the world before the electronic hearth. Just as rapid transit, especially air travel, can be perceived as having made the world a smaller place, wireless transmission of images in the form of television signals can be argued to have shrunk the world, perhaps making space disappear completely as a significant factor in our lives. But, just like rapid transit, the influence of electronic media on our perception of space is more complex than this. Spaces are connected to one another using sets of rules that have more to do with politics, power, and preference than with physics. When Marshall McLuhan, a pioneering Canadian thinker in media studies and author of the influential slogan "the medium is the message," described the impact of new media as having converted the world into a kind of "global village," this is precisely the kind of transformation in the use of space that he meant. Like villagers, we form allegiances, links, and unions with other individuals, but the far reach of invisible waves makes physical distance irrelevant to the formation of these connections.

The mind-warping power of television has in some ways worked in concert with other space-devouring developments of the past century. In previous chapters, we saw how the designs of suburban homes and neighborhoods have discouraged us from venturing outside the front door. Monofunctional designs have given us few real destinations, and the quiet streets that result have produced a social isolation and physical torpor that prevents the growth of healthy minds and bodies. The availability of a constant stream of images from across the globe has given us a much-craved window on the world that is intelligible to us mostly because of our mind's penchant for the virtual. We find it so easy to slip and slide through space that there is little effort involved in our letting the two-dimensional images that flicker across our living rooms stand in place of real life. Though the medium of television has not lived up to its early utopian promise to educate generations of cosmo-politan, connected human beings, this is less an intrinsic failure of the medium itself and more, perhaps, a failure of our own imagi-nations mixed in with a good amount of greed. It didn't take long to discover that televised images were incredibly powerful tools to entice us to buy things, and, public broadcasting notwithstanding, this has been the main use of television throughout most of its almost eighty-year history.

——————

It's almost a shock to recognize that the Internet is less than twenty years old. Given its deep penetration into almost every aspect of our lives, it is hard to imagine how we ever managed to limp along without it.

My first use of the Internet for anything other than to read in newsgroups about the future promise of the Internet was to settle a debate that I was having with my wife about the rules of cribbage. I entered "cribbage" into the rudimentary search engine of the day (does anyone remember Gopher?), waited for what would by current standards be an intolerably long time—long enough now to download a full novel—and there, to my astonishment, sat before me on the screen a simple text document outlining all rules of cribbage. The only thing that fascinated me more than that the search engine had been able to turn up this document from somewhere within the network was that somebody had bothered to write the document and to make it available to all.

Contrast my early fumbling efforts with the present-day deep penetration of Wikipedia, Google searches, and YouTube in our home and even our working lives. Anyone within reach of an Internet portal has at their fingertips an astonishing mass of text, image, and sound that can be delivered to their screen in milliseconds.

Radio and television networks are mostly owned by large economic stakeholders who maintain a firm grip on what is and is not presented to their audiences. The Internet, at least for now, is not only accessible from almost anywhere in the world but also simple enough that any garden-variety user can post content. Even when repressive governments attempt to control what content is available to their citizens, clever users quickly find ways to skirt around such restrictions (witness the recent efforts of Internet users to publicize the efforts of the Chinese government to quash unrest in Tibet). It is this interactivity and collective ownership of the content of the Internet that so distinguishes it from other types of mass media, and these are also the features that fill it with potential to have a positive transformative effect on our relationship with physical space.

In a way, the Internet poses the same space-squashing conundrum as any other form of electronic communication. The enormous expenditures of energy that underpin the computers, servers, fiber-optic cables, air-conditioning, cooling fans, and service vehicles driving around on freeways to repair chewed wires and broken switches enable us to be completely shielded from the real distances that are involved when we press a button on a keyboard and suddenly find ourselves looking through a live webcam at a lion dozing in the sun in Kruger National Park.

In fact, navigation from website to website by a series of clicks mirrors the way that our mind processes space. Internet sites are connected to one another as nodes in a topology. When we click on a link, we normally have no idea of the distances or directions that separate the sources of the sites that we're viewing, and nor do we care about them. This is just a more extreme version of our tendency to collapse the geometry of space to a simple topology as we navigate our way through our everyday lives, and it may be one of the reasons why point-and-click Web browsers are so intuitive and easy for us to learn to use. I've watched my 80-year-old father and my two-year-old son master browsers with equal facility, and, like everyone else, I've seen them become tangled in a thicket of distracting visions, losing track of where they started and where they were trying to go. (Tech watchers have an official name for this behavior: WWILFing, for "What was I looking for?") This ability of websites to engulf and "lose" us in content is even something that Web designers strive to include in a site. Just as casino designers or shopping mall architects might try to build physical spaces that make it easy for us to enter and hard to leave, cyber-architects have as much social control (and presumably economic benefit to gain) when they tap into the way that our minds compute virtual spaces.

The main difference between the Internet and more passive electronic media is its capability for personal interactivity. Though many websites are set up to be simple digital versions of print media (some newspaper websites are a good example), others are designed to encourage users to willfully and actively navigate to different parts of a site, building the shape of their overall experience on the fly. Just as we may wander from room to room in a gallery to gaze at the offerings on display, we can do the same kind of thing on the Internet by choosing which parts of a site to visit and how long to linger. Many sites also allow us to structure our experiences according to more complex user inputs than simple mouse clicks on links. We might be asked to answer questions, for example, and our answers might be used to steer us on a particular path through the content of a site. So although electronic communication via computer networks distorts space in the same way as any other technology that allows us to move our virtual selves at light speed from one place to another, there is much more potential for involving the user in a richer mental experience than simply gazing at a centrally controlled flow of images. This means that there are real prospects for using this technology to effect positive changes to human mental states, including our conceptions of the organization of space and the connections between ourselves and other parts of the world. These prospects have been enhanced considerably in the last decade by the widespread availability of geocoded data on the Internet.

GEO-EVERYTHING

In 1995, the United States announced the completion of a remarkable project that had evolved slowly over the previous three decades. The deployment of 24 space satellites composing what was called the NAVSTAR Global Positioning System made it possible for an electronic device on the surface of the earth to calculate its location to within a few meters.[5] One of the main motivations for the system was its envisioned military uses, and for the first few years of its existence, an error was deliberately introduced into the GPS signals offered to civilians so that these signals could be accurate only to within about 30 meters. During the Gulf War, a shortage of military-grade GPS receivers prompted the U.S. military to remove this source of error so that more widely available commercial GPS receivers, bought mostly by boaters and hunters, could be used on the battlefield. Since that time, the distorting error has remained inactive, and GPS signals have become a mainstay of air and marine navigation. Indeed, though the military retains the capability of blocking the GPS signals from reaching any part of the globe, to do so would probably lead to catastrophe. Today, the evolution of remarkably accurate and tiny GPS receiver chips (about the size of a pinkie fingernail) has meant that these signals are widely available in a variety of consumer products, including laptop computers, pocket computers, cell phones, cars, and digital cameras. The wide availability of GPS-enabled devices, along with geographic software that is useful for professionals but user friendly enough for casual users, has led to sharp interest in tagging objects with information about location. Google's stunning free software Google Earth, which enables users to view everything from a snapshot of the entire globe to detailed street-level views of major urban centers, has led to a craze for what some refer to as "geo-everything."

At first blush, geo-coding, in which our activities, snapshots, phone calls, and blogs are tagged with precise latitude and longitude information, might seem like something that would interest only a thin segment of the technophile population, but there is something compelling about connecting the moments, thoughts, and glimpses of our lives to actual rock-and-brick locations. Is it possible that technology is slowly but surely helping us to turn full circle and to find ingenious ways to reconnect ourselves to physical place?

There are countless practical reasons for the public's fascination with geo-coded data. It's clever to be able to reconstruct a vacation by opening a beautiful cartographic image on a computer that is overlaid with catalogs of photographs. It's handy to be able to dial up a map of the 10 closest coffee shops to one's home, with a personal ranking of the quality of the java on offer at each of them. But beyond the practicalities, being able to tag our experiences with a carefully measured *where*, especially if it is beyond our dulled senses to pick up location more directly, seems to satisfy a deep craving for place that we'd forgotten we had.

One of many interesting developments in the use of the Internet to connect us to one another came in the form of social networking. Some applications, Facebook being perhaps the best known of them, allow users to sign up for free accounts and post messages to virtual locations that can be read either by anyone who stumbles across their site or only by those who have been explicitly selected as friends, depending on the privacy level that is chosen. Social networking sites allow users to share messages, images, sounds, or really any kind of digital content. What such social networking applications accomplish is that they make each user a kind of node in a network, and permit users to make their connections to other nodes transparent and readable either to the general public or to a select few. Though social networking applications at first appealed to a small segment of the population—mostly university students trying to keep track of rapidly evolving social lives not always tied to physical places—their popularity has skyrocketed with the general public. The possibilities for wider applications, including some marketing of products, and the availability of much custom software designed to enhance the features of social networking sites, are now beginning to come to fruition.

Some social networking applications offer the added feature of location-based service, usually designed to allow one to post brief messages from mobile devices. These messages can be either location coded by hand ("I am now in my kitchen making dinner") or automatically geo-coded using GPS data. A quick browse through a representative sample of public messages posted using such applications as Twitter (which is not location coded) or Brightkite (which is) suggests that most users post very ordinary messages about the mundane details of their daily lives—hardly the stuff of a Tolstoy novel. But to discard the potential of such virtual networks based on current usage would perhaps be to make the same mistake that I made about 15 years ago when I assumed the Internet was useful only for looking up the rules of card games.[6]

One interesting twist that will soon be more generally available will allow your networked friends to access continuous information about your spatial location based on an encoded GPS signal. This will make it possible to, for example, set up proximity alerts that will inform you of the nearby presence of friends, creditors, and perhaps even ex-lovers. Such capabilities will engender new ways in which technology can produce electronic proxies for our understanding of geographic space.

In a way, computer-based social networking can be considered the modern equivalent of ancient tracking methods used by desert Bedouins riding camels or Inuit hunters crossing the land on dogsleds. The main difference is that the skill set required of users has changed considerably. Whereas an ancient tracker might need to learn how to read camel dung or dog footprints to extract social history, his modern equivalent is able to access such information directly. The key to both ancient and modern social wayfaring is that both geo-coders and Inuit hunters are engaged in a pursuit that is helping them to connect a network of human activity to a physical landscape. The connection of narrative to geography is a time-tested method for throwing our connections to places into sharp relief.

THE DEEP GREEN CALMNESS OF UBIQUITY

GPS receivers are only one example of compact computing devices that give us mobile access to significant information about our relationship to the world. Sensors can also read other environmental variables such as pressure, temperature, wind, UV radiation, air quality, and noise. In addition, more complicated derived variables (stock market trends, approaching weather systems, road traffic patterns) can be streamed to remote devices via wireless Internet connections. All such sensors and devices could be deployed throughout our environment to shroud us in a kind of distributed intelligence as we went about our daily business. This approach to embedding intelligence in our environment, termed "ubiquitous computing" by techno-pioneer Mark Weiser, is the inverse of virtual reality. Whereas the intention of a virtual environment is to produce a painstakingly detailed simulation of a real environment that engages all of the user's sensory systems, the ubiquitous computing approach throws the intelligence offered by computer designers out into the world, relegating even desktop computing to more of a background role.

In a brilliant exposition of the potential of ubiquitous computing to revolutionize our relationship with our environment,

Weiser, along with his collaborator, John Seely Brown, described how computing devices could fill our lives with what they described as "calm technology."[7] In contrast to focal computing devices like desktop computers or BlackBerrys that demand our full attention, calm technologies work in the background, engaging the periphery of our senses to inform us gently about states of the environment that might not otherwise be obvious to us. One low-tech example of a calm technology is the inner-office window that connects the occupant of an office with the outside world, the people walking past in the hall, the intended visitor who has peeked in a few times, waiting for a chance to enter. The window doesn't fully engage focal attention, so it doesn't enrage or annoy, but it does provide an ambient source of information about goings-on in the world outside the room.

Weiser and Seely Brown describe another beautiful example of calm technology that comes in the form of an artistic installation piece by Natalie Jeremijenko called *The Dangling String*. The piece consists of a 2.5-meter length of plastic string that is connected at one end to a small motor. The motor receives inputs from an Internet connection. The vigor of the string's movement, ranging from a gentle wave to a frenzied dance, correlates with the amount of network activity. Ambient Technologies, an offshoot of MIT's Machines That Think laboratory devoted to ubiquitous computing initiatives, has developed some products that embody calm technology principles. The Ambient Orb, for example, is a spherical light whose glowing color can be made to correlate with virtually any value that can be streamed from the Internet, such as air quality, pollen counts, or stock tickers.[8] An orb placed in a room can provide us with a gentle background monitor of some useful piece of information that never demands our full attention but helps to fill in a greater context and can potentially ground us to the larger environment.

What I find most interesting about Weiser and Seely Brown's argument for calm technology is that, in many ways, the language they use to describe their approach to embedding intelligence in the design of everyday settings is much in accord with the language used to explain how natural environments can be attractive and restorative to the human psyche. In explaining the attraction of devices that inform gently from the sidelines rather than blaring headlines from front and center, Weiser and Seely Brown say, "The periphery connects us effortlessly to a myriad of familiar details. This connection to the world we called 'locatedness,' and it is the fundamental gift that the periphery gives us."[9] Is it possible that the artful deployment of technology in our everyday settings will not only help us to connect ourselves to places but will also help to ease the attentional overloads produced by chronic exposure to busy and chaotic urban settings?

VIRTUALIZING THE REAL AND REALIZING THE VIRTUAL

Advocates of ubiquitous computing approaches have been quick to dissociate themselves from the world of virtual reality, but it is hard to envision a technological future that does not include some measure of virtual life. In a way, whenever we talk on a cell phone, thumb a message on a handheld device, or listen to an iPod, we are producing a fracture between our physical and our virtual locations. Though architecture and appliances that inform us of the ambient state of the wider world, or even respond to our individual moods as we enter or move close to them, may help to bring our focus back into the here and now, our unique cognitive architecture, tailor-made for leaping continents with a single mental bound, would be impossible to rein in completely. A better approach might be to find fruitful ways to combine our virtual worlds with a physical reality studded with ambient computer intelligence into a rich form of mixed reality.

Recently, providers of the Second Life metaverse, in yet another risky move, decided to go "open." Stripped of cyber-speak, this means that Linden Labs has made available to the public the computer code that generates the software that allows Second Life to run on a personal computer. This is no small thing. What it means is that knowledgeable programmers can enhance the Second Life software to bring additional features to it. One possibility that occurred to many groups as soon as the decision to open the software was made was that Second Life and first life—the real and not the virtual one—could be brought into closer contact than ever before by employing mixed reality designs.

What if a user's behavior in the real world could, through an appropriate interface, be reflected in the virtual one? If the output of a skin-temperature monitor attached to a user could be jacked into the metaverse, perhaps the user's avatar could be made to blush. If a weather-monitoring station's wind speed measurements could be entered into Second Life, perhaps virtual leaves could shimmer in real breezes and virtual tree limbs could bend.

The unanswered question is whether such a convergence of the real and the virtual would exert an effect on the sensibilities of the user. If we could visit a simulated location in Second Life, say a boiling desert in Northern Africa, and witness the effect of an overheating climate on local events firsthand through the eyes of an avatar, would it increase our understanding of the connection between drought, famine, and starvation and the level at which we set our home air-conditioners? Making such connections would be difficult for all of the reasons discussed earlier. While hyper-voyages through the Ethernet remain a sporadically experienced curiosity available only to those with the money and the technological savvy to carry them off, they will be amusing, perhaps even informative, but hardly transformative. In fact, such crazy-quilt electronic adventures may only serve to exacerbate the frailty of our connections to real space by offering all places to us in an instant, all on an equal footing, a horizontal electronic plain in which physical distance was as meaningless to us as it ever has been in the realm of rapid communication.

Although combining real-life events with virtual ones within a desktop environment offers many interesting opportunities for us to understand our connections to the planet, the onus would be on us to deliberately seek out such experiences and to do a considerable amount of work to relate what we experience to our own lives in any sensible way. A more promising approach would be to embed pieces of virtual life into our everyday experiences. If, on my walk to work every day, I must pass a solid sculpture in a park that renders the quantity of remaining sea ice in a tangible and direct way by varying its size and shape, then a fact of a remote environment will, over time, worm itself into my consciousness and become a part of my inner landscape. In the manner of the best in ambient computing, the sea ice indicator will not be in my face as a blaring beacon of guilt but will form a part of the deep background of my geographic experience as I walk a familiar route.

In our Research Laboratory for Immersive Virtual Environments (RELIVE) at the University of Waterloo, we build richly detailed simulations of buildings and landscapes that rival those offered by less immersive environments like Second Life.[10] The models begin with a skeleton of the positions of the main features of a virtual world, but fine details of visual texture and lighting can be added to generate compelling illusions of real spaces. Participants in our experiments view these simulations by wearing headsets that contain small screens much like those used on laptop computers.

Specialized sensors keep track of the user's movements as they walk around the room, allowing computers to update the images that appear on the screens so that users feel completely immersed in the simulation. Virtual reality researchers speak of an ephemeral quality called presence, by which they mean the quality of a virtual experience that leads one to believe at some level that one is immersed in a real space. Though nobody could really forget that these virtual environments are not the same as physical space, the best of such environments can produce a remarkable depth of presence. One virtual reality designer has produced a simulation of a car so realistic that he cannot persuade himself to drive this car over a virtual cliff. In our laboratory, we can generate a convincing illusion of a deep pit that suddenly appears in the floor of an ordinary-looking art gallery. Though we can coax ourselves to walk to the edge of the pit and look down, even experienced users walk gingerly, and physiological monitoring equipment betrays our rapid heart rates and sweaty palms if we force ourselves to "walk the plank" across the pit.

So what happens when people are placed into such virtual spaces? In addition to feelings of anxiety and arousal when challenged to drive over cliffs or jump into chasms, we are surprisingly quick to accept such alternative universes as the real thing. There are many informal signs of such deep immersion. For one thing, virtual navigators soon forget that real, concrete walls lie outside the realm of their helmets, and we must guard their safety closely to prevent them from banging into the walls. In addition, after users have worn the helmet for a few minutes, removing it gives rise to a "popping" sensation as they discover and remember the real world that they had just left. It's difficult to describe in words the relationship between the physical and virtual worlds experienced by users who are jacked in to our equipment. In a way, it is like a dream state.

Some of us, when we dream, become aware that we are dreaming and so, for a fleeting time, can be simultaneously aware of our dream selves involved in a narrative and of ourselves as sleeping bodies lying in bed. Those who have "willed" themselves out of a precarious situation in a nightmare will be familiar with this sensation. Some have been able to cultivate this kind of dual awareness, or "lucid dreaming," perhaps even finding ways to manipulate dream content as it takes place. The virtual reality experience is a little like this sensation. Though the graphic worlds we see are of very high quality, we can never forget completely that they are embedded in a real-world context.

In many of our studies, we stand beside our volunteers and talk to them as they work in virtual space. It is not hard to move one's sense of self back and forth between the physical room containing the experimenter's voice and a virtual realm filled with parks, trees, buildings, and avatars, but, as yet, nobody knows what consequences might ensue from longer-term chronic exposure to highly immersive virtual settings. This is a very important question, for such technology, so far mostly confined to research centers, is clearly destined for people's homes in a few short years.

More careful measures of behavior in virtual environments suggest that people tend to move and linger through our digital spaces in much the same way as in real spaces, and that the rules of space syntax that work so well to predict movement through buildings and cities appear to hold up in virtual spaces as well. Most interesting, the same disregard for angle and distance that can confuse residents of real space also applies in virtual spaces, and it can be used to interesting effect. In one study conducted at a large virtual reality laboratory at Brown University in Providence, researchers produced a large virtual maze that contained what they called wormholes. These holes were virtual portals that catapulted observers from one position in the maze to another remote location.[11] To get some idea of the effect, imagine walking out the front door of your house and discovering yourself back inside the house, somewhere near the back door. The simulation was designed to conceal the presence of the wormholes from the virtual travelers. As the whole space was unfamiliar to them, and one area looked much like another, this was not difficult to arrange. In navigation tests, not only did the volunteers fail to notice the wormhole effect at all but their routes through the virtual space suggested that they had little understanding of the connections between the arrangement of these holes and the real spatial relationships of different parts of the maze. When asked to find shortcuts from one location in the maze to another, the routes they chose suggested that they had been oblivious to the geometry of the room, computing location and path using topology alone.

Much of our research at RELIVE relates to the testing of principles involved in the design of built spaces in houses, buildings, and cities. A skillful designer can produce a model of a proposed building or streetscape that can then be presented to users to measure their reactions. Not only can we ask participants whether they like what they see but we can monitor their movements to see which parts of our digital spaces attract their eyes and then their feet. All the while, sensors record the reactions of their bodily systems to what they experience. Such systems can be used powerfully to understand how human minds process space, but they can also help solve interesting practical problems. Thomas Seebohm, an architect at the University of Waterloo's School of Architecture and a pioneer in the use of such visualization methods in the design process, used large-scale detailed simulations of entire city blocks to underpin the public consultation process in urban-planning decisions. Though drawings and tabletop scale models are of some help in letting people imagine how construction projects might transform their lived spaces, more immersive environments that allow people to walk through detailed virtual mock-ups of a new building or development project can be used to provide a much more comprehensive vision of alternative futures.

In our laboratory, we are also designing virtual structures whose size and shape adapt over time to reflect the preferences and interest of the observer, as measured by their movements and their physiological state. This symbiotic dance of human user and virtual space should help us to define and optimize the shape of dwellings and workplaces, and might also help us to understand how different types of spaces can be adapted to individual preferences. Another local architect and artist, Philip Beesley, is one of a group of individuals with an interest in what has been called responsive architecture, in which buildings might sense the movements and perhaps even the physiology of their occupants, and adjust their properties accordingly to yield maximal comfort.[12] On a simple level this could involve such things as adjusting ambient temperature to coordinate with the body temperature of an occupant, but it might also be possible to adjust the shape and appearance of space to enshroud the user in a zone of comfort, ease, and security.

———————

High-powered graphics, surround sound, and compelling narratives can take an occupant of a virtual place a long way away from the real world, but what can make such digital constructions crackle with life is the presence of other beings. In another demonstration of Jane Jacobs's "life attracts life" dictum that even she might never have imagined, one of the surest ways to boost a feeling of presence in a virtual world is to share that world with other people. This is one reason for the huge success of efforts like Second Life and

World of Warcraft. What such metaverses lack in immersive realism they make up for by being social places in which one can walk up to other beings for conversation, laughter, dancing, dating, hugging, and even a romp in the bedroom if the virtual stars are just right. These opportunities for social interaction can so draw a user into a metaverse that they may sometimes feel they never want to leave it, even though their entire embodiment is invested in a few hundred colorful pixels on a small screen on a laptop. Now imagine that such shared presence can be enjoyed by users of a more powerful, more immersive experience such as the one provided by helmet-based virtual reality, in which users wear a headpiece containing a visual display and a pair of headphones that essentially "lock" them into the virtual world and exclude the actual physical world that surrounds them.

At Waterloo's RELIVE, along with a few other such installations, two users in different physical locations can don headsets and meet one another in a shared virtual space. Like Second Life, users meet one another's virtual embodiments in the form of digital avatars, but these avatars can be rendered in incredible detail with realistic skin textures, clothing, and gestures. When the users wear additional sensors (such as special gloves that allow the main computers to track the positions of hand and fingers), their digital avatars can interact physically with one another. And though the technology for truly realistic bodily interactions (such as handshakes) is still very much in development, it is possible to outfit such sensors with some degree of feedback so that an occupant of a shared virtual space can not only "reach out and touch someone" but both those doing the touching and those who are touched actually feel a physical sensation on their skin that corresponds to the virtual action.

If such technology was to be made widely available, the consequences for social interaction are simply mind boggling. For example, teleconferencing has been touted as a means of improving the efficiency of communications by allowing a group of individuals to meet in a shared auditory space. But anyone who has ever participated in a conference call knows that when the only information available consists of words, pauses, and tone of voice, it can be difficult to keep track of the thread of conversation, let alone to leverage the normal dynamics of a group experience to promote a clear set of actions. There is no better way to make one appreciate the complexities of group interactions than to reduce everyone to disembodied voices. Adding a video link is scarcely better, as the effect is really just to add to the melange of voices a set of slightly dynamic headshots of people sitting at tables. Natural movement, body language, and eye contact are entirely missing.

Now imagine sitting in a room that exists only in the shared mind spaces of the members of the meeting. It is possible not only to see three-dimensional embodiments of other members of the group but to talk to them, watch their gestures, facial expressions, and patterns of eye movements, and make eye contact. Though there are some technical hurdles left to work out to make such meetings a common reality (the biggest one perhaps being Internet connections of sufficient bandwidth to allow the real-time transfer of data from one place to another), many of the most difficult challenges have been overcome.

The widespread possibility of such virtual interchanges could revolutionize many business practices, and the use of immersive environments in conjunction with social networking staggers the mind. Imagine being able to go to a souped-up version of an Internet café where you could convene a meeting of friends scattered across the globe, sit in a circle, converse, share music or a movie, and perhaps even hold hands and hug. In a sense, nothing that I have just described is particularly new. People have been going on cyber-dates for years, and there is plenty of evidence that even text-based cyber-sex using an email program can be compelling enough for people to leave stable marriages to take up with new partners whom they have never met in the flesh. New, though, is that such interchanges will be possible using technology that will provide for undreamt-of levels of realism and presence.

Now go one step further, and remember who we are and where we have come from. When we send our avatars down fiber-optic cables to meet up with avatars representing friends, business partners, political counterparts, or potential lovers, what guarantee is there that we must convey accurate representations of ourselves? Everyone knows of people who can woo with words, make love to a camera, or otherwise convey enhanced versions of themselves that are, in some way, larger than life. How much more potential for interpersonal havoc might there be when such "transformed social interactions," as psychologists have called them, can be carried out by using technology to manipulate the features of lifelike, solid-looking avatars who might even possess warm virtual flesh and a beating virtual heart? Again, this is not a particularly new question. All our experiences are mediated by something, whether it is text, image, or just the distorting channels of our own senses, cognitions, and emotional biases. Jim Blascovich, a pioneer of the use of virtual reality to study human interactions, puts it simply. "Everything is virtual," he says.[13] But what is new is the extent to which it will be possible to recruit sophisticated technology to plant false beliefs in others using not only word and image but movement, body language, facial expression, and something approaching flesh-to-flesh contact in a virtual environment. Once such technology is more widely available, Pandora will be out of the box, and past experience suggests that any attempt at regulation will be fraught with difficulty and will probably fail. It may be that our best hope of preventing such technology from eroding the quality of genuine physical human interaction is to act now to study and understand both the scope and the limits of its virtual counterpart.

Although equipment that can produce deeply immersive experiences of interactions with virtual partners is likely to be confined to specialized facilities for the foreseeable future, improvements in computer graphics power and speed are rapidly bringing decent-quality immersive virtual reality experiences to within the reach of consumer budgets. A baseline system for generating helmet-based digital worlds with some motion tracking can be had for about $20,000, around the same price as a high-end home theater. The much more affordable Nintendo Wii, with its wireless controller, is perhaps the thin edge of the wedge for consumer-grade virtual reality environments meant to be installed in homes. This controller gives users the ability to immerse themselves physically in a script by holding a wand whose movements through three-dimensional space are incorporated into such things as virtual tennis, golf, and combat. Like Second Life and similar ventures, we adapt so well to developments such as these because of our extremely flexible context- and view-based understanding of physical space. Our minds leap joyfully from one kind of space to another with scarcely a backward glance because they are built to absorb the shocks of a world put together one brief, disconnected glimpse at a time.

The advent of such technologies in our homes presents both great opportunities and great risks. If a game like StarCraft can cause players to forget to feed themselves or to go for days without sleep, then what happens when we can invite Attila the Hun into our living room to engage in swordplay with a weapon whose heft we feel in our hands? At the very least, we will need to come to a fuller understanding of the elements of game play that contribute to addiction. But we will also have to take seriously the issue of how to ensure that the lines between physical and virtual reality are drawn cleanly enough to prevent social havoc.

Understanding more about how human beings, especially young ones, engage with immersive virtual worlds could help us to arrive at a rational rating system for regulating the sales of such software and educating users about the potential risks. As virtual environments become more common and compelling, we will need to address questions about the extent to which we should expect normal social mores and legal frameworks to apply in the virtual realm.

Such questions are already being raised in Linden Lab's Second Life. When the world was initially set up, residents who refused to adhere to a code of conduct that included provisions designed to prevent harassment of other members were incarcerated in "the corn field," a dark corner of a simulated farmer's field in which nothing happened. Avatars were sentenced to various lengths of forced stays in the field. More recently, a lawyer has sued Linden Labs for violating trade practices in a land deal where the land in question consisted of simulated real estate in Second Life. Though this might be seen as a publicity stunt, such issues will arise more frequently as we begin to use virtual realms as legitimate extensions of our lived spaces and invest time, emotion, and, perhaps most important, money in our virtual lives. Imagine how these kinds of issues could multiply in virtual realms in which we can see, hear, and feel one another with much higher degrees of fidelity than are currently possible.

In spite of the challenges and risks, wider availability of good-quality immersive virtual experiences in homes holds great promise. Educators will have tremendous opportunities to take advantage of an inherent appetite for game play in children, providing them with knowledge and experience that would be hard to acquire in any other way. Imagine a game like the hugely popular geography quest

"Where in the World is Carmen Sandiego?" rendered in full immersion and in three dimensions. In institutions such as universities and perhaps larger community centers, more powerful immersive virtual reality setups could generate precisely detailed walkthroughs of ancient Rome or Shakespeare's Globe Theatre in which students could learn by talking to local "residents," participating in group activities, and manipulating virtual objects with the look and feel of the real thing. Applications of such technologies are limited only by the imagination of the content developer.

VIRTUAL DYSTOPIAS

Though we have, perhaps, never been so close to realizing the technology to make it so, the idea that we can embed our minds into sweeping simulations of space to produce compelling illusions of reality is far from new. The earliest versions of such simulated realities, such as Morton Heilig's Sensorama, were based on unwieldy mechanical contraptions that did a credible job of lifting the focus of a person's embodiment from their physical place to a virtual one.[14] In the 1980s, when virtual reality helmets and computer technology began to reach the minimal technical standard to have some industrial uses, there was tremendous media interest in the possibility that virtual reality would enter the mainstream as a means of entertainment and communication. Though simulation methods are now commonly used in industry and in the military, the predicted mass consumer rush to this new technology never took place. Many of the reasons for this have been technological—there is still a large gap between what can be shown on tiny screens in front of the eyes and what the real world has to offer. Virtual reality also suffered seriously from overhype. The experience in a virtual reality arcade game at an amusement park fell so far short of expectations generated by the first flush of media reports about this emerging technology that the public lost interest quickly.

In addition to the practical limitations posed by the problem of presenting consumers with decent-quality immersive experiences in virtual worlds, certain psychological factors were at play, and continue to lurk in the popular consciousness. William Gibson's novel *Neuromancer,* Neal Stephenson's *Snow Crash,* and the popular series of *Matrix* movies by the Wachowski brothers all present bleak visions of a future in which technology allows us to build virtual spaces that are indistinguishable from physical ones. In each of these cases, and in many others, we are given glimpses of dystopic worlds in which parallel virtual universes are used like weapons to produce mass delusions that crush the human spirit, or in which the virtual worlds we create become forums for the exercise of our basest impulses, untrammeled by the normal mores of social conduct and even freed from the operation of physical law. Optimistic visions of how virtual realities might enhance our lives are remarkably rare. In fact, even in Second Life, a low-immersion virtual space that is being marketed aggressively by mainstream media as the harbinger of a new way of using virtual embodiment to communicate, have fun, and do business, a seamy side appears to be emerging. Children must be sequestered on their own servers with stringent controls lest they be stalked. In a simulation of a Darfur refugee camp designed to raise consciousness of a real-world horror, self-styled superhero vigilantes have had to organize themselves to protect the camp from raids by apparent racist groups intent on vandalism and hateful graffiti.[15]

Though his pungent criticism of the effects of rapid information transmission transcends virtual reality and includes any form of communication that collapses space, French philosopher Paul Virilio warns us forebodingly of the impact on human relations and power politics when all space collapses to a single point and when everything happens simultaneously.[16] One of the greatest consequences of such technologies, already well under way, is that the power of one group over another will come to be dominated by what we can see rather than by where we are.

The hegemony of glimpses can be seen in the way wars are now conducted. Whenever possible, battles are conducted from the air. Occupying a territory can often mean being able to see it from overhead, by satellite or by high-altitude drone that can aim cruise missiles, or via gigantic surveillance planes such as the AWACS. Though bombs may still drop from planes, often the *threat* of bombs will be enough to control territory. Because our eyes can be placed anywhere in an instant, all walls have dropped and we can both possess and be possessed by sight alone.

This might not seem like an entirely negative point of view. A vision of warfare with soldiers writhing together in mud and blood could be replaced by one composed of nothing more than dueling electronic eyes, but it must be remembered that those few who control the stares will still command the lives of the masses of people living on the planetary surface. The main difference will be that power relationships, like fleeting rainbows in the sky, can be cast and recast in an instant. The dystopic vision of our future is one in which all 6 billion of us are trapped huddling on a tiny, globe-sized dot of space, tangled up among the views of a mass of powerful, all-seeing eyes that control our fate remotely and instantaneously, but without ever really touching us.

The dystopic visions of philosophers, artists, filmmakers, and writers should not necessarily be construed as dooming us to a future filled with techno-gloomy realpolitik, but there is no question that the technologies I have discussed in this chapter, along with many others that are on the way (quantum computing and nanotechnology, to name two), have the potential to produce revolutionary changes in our lives. Many of these developments were possible not just because our species has a clever mind for inventing gadgets but also because we possess a perceptual and cognitive architecture that has so far found ways to cope with the fragmentations of space that those gadgets have brought about.

Understanding how we are affected by these transformations in how we live in space is perhaps no less urgent than the challenges presented by climate change. We know from our history that, all calls for prudent forethought notwithstanding, whatever we *can* make, we *will* make. The onus is on today's bright thinkers in science, the humanities, and the arts to try to anticipate and influence for the better the products of our digital constructions.

CHAPTER 11
GREENSPACE

How the Features of Our Spatial Brain
Influence Our Connections to, and Neglect of,
Our Natural Environment

I am at two with nature.

WOODY ALLEN

As we crossed a parking lot, my wife and I watched the driver of a parked car roll down his window, toss the wrappers from his fast-food lunch onto the pavement, and then close his window again. Karen, in the kind of gesture that I have come to know very well and to love without limit (in spite of the constant risk that she will cause me to have my head broken), picked up the pile of trash and rapped on his window. When he lowered it, she passed the bag back to him and asked him to take it with him or to find a garbage bin to put it in. He was knocked off balance, which may be what saved me from a broken nose. He took back the bag without comment and drove away. I'm not sure how much time he spent thinking about this episode, or whether he tossed his garbage out the window again as soon as he had turned the corner. But I do know that this simple transaction preoccupied Karen and me for quite some time. I still think about it and tell people the story, because I think the man's actions have a larger meaning. It would be easy to write him off as an asocial ignoramus, but there is much evidence that what we saw was not an isolated or unprecedented incident.

We've all seen garbage strewn along the edges of highways and in the gutters of city streets. Although we might like to think that some of this mess is accidental—papers blow out of doors and windows of vehicles, garbage cans are upset by wind or animals—we know that at least some of it appears because people deliberately throw it on the ground. Where do they think this garbage goes? It may be that they just don't think about it at all. Such behavior is a small sign of the way that we mentally divide spaces into inside spaces and outside spaces, treating the boundaries between the two as though they are absolute and impossible to breach. How different is my own behavior when I toss a bag of trash by the side of the curb without a second thought? We have a broad expectation that such things will be "taken care of," where as far as we're concerned, this "care" is more a matter of "out of sight, out of mind."

Modern Western houses, with their steel, dead-bolted doors and thermal windows, may be a recent invention beyond the reach of most of the world's population, but efforts to build dwellings that enclose, separate, and protect us are as old as any human fossilized remains that we have ever found. Though the walls of our houses are intended to protect us from harsh climate and to afford some privacy, they also have the consequence of erecting an impermeable mental barrier between the interior and the exterior spaces of our world.

As the planet careens toward environmental catastrophe, we are bombarded daily with messages designed to wake us up to the imminent risk that we may exterminate ourselves, along with most other life on the planet. Yet when we hear the word *environment,* we are overwhelmingly likely to think of a natural setting—a forest, a meadow, a range of mountains. In our minds, there is such a cleft between these exterior spaces and the interiors of our homes, offices, and factories that it interferes with our ability to appreciate the urgency of our situation for what it is. Just as the fellow in the parking lot was able to toss his trash out of his car window and excise the problem from his own mind, we behave as though the "problem" of the environment affects only those pristine outdoor settings and has no bearing on the interiors in which we spend most of our lives. Because spaces are completely separated by enclosures, we have difficulty connecting the warm security of our living rooms with the toxic foam floating down a river in the parkland just outside our doors.

I believe that our inability to make connections between different types of space—the indoors and the outdoors, the urban and the rural—has a basis in the makeup of the human mind and the way that we engage with space. We handle the immediately visible spaces before us very well, but our mental understanding of how those spaces are connected to the larger realms beyond our purview is fuzzy at best.

Concerns about the state of the environment have been with us for a long time. When I was growing up in the 1960s, I well remember the alarming descriptions of air pollution and the devastation of the waters of the Great Lakes in North America. It seemed as though every newscast brought messages of doom and gloom sufficiently urgent that I wondered whether I would survive to adulthood or be choked on exhaust fumes and then washed away in a sea of oily lake water. Organizations such as Pollution Probe gave shrill warnings that unless we started to fix things *now,* the planet would soon be uninhabitable. At the same time, and just as vividly, I remember the soothing reassurances of many of the adults in my life, who told me that "science" would find answers to these problems and that, ultimately, we would all be living underneath giant glass domes, hermetically sealed from the "outside" no matter what devastations might have been wrought there by the errors of the past. In other words, the ultimate solution to environmental destruction would be to shift the lines between inside and outside so that the ambit of our thermostatically controlled living rooms would be extended outward to encompass the nice parks that we would need to play soccer and to have picnics. The world may be going to hell in a handcart, but moving the walls essentially redefines the world for us and so allows us to maintain our current habits and way of life, regardless of what might be happening outside our safe domes.

WALLING OFF NATURE USING SPACE AND TIME

It isn't just the enclosing walls of the built environment that make us feel separated from the natural world. Distortions in the natural relationships between human movement and the scale and aspect of space can also rend the connections between our mind and biological space. When Bruce Chatwin sat in a fast-moving car beside his Australian Aboriginal friend and listened to him try to warble a speeded-up version of a songline that had evolved over thousands of years to match landscape to a man walking through the wilderness, he was hearing a remarkable illustration of the breakdown of the integral relationship between our own movements and our conceptions of space.[1] Though our large cities and fast-paced way of life may make them necessary, modern methods of rapid transportation have completely changed the meaning of space. Now, rather than being the main way that we get from one place to another, walking has become a sporadic event, a small punctuation used mostly to convey us from one moving machine to another. We can cope well with these staccato transitions from place to place only because the structure of our mind makes it easy for us to relinquish our grasp on the metric of space, but the distortions that result from such quick movements serve to increase the fragility of our understanding of spatial connections between different parts of the world.

I often travel by airplane with my children. Recently, I've noticed an interesting thing about my toddlers' experience of flight. As far as I can tell, they have no clear understanding of the purpose of the airplane other than that it is a machine that makes a huge amount of noise and traps them in a space that is slightly too small for their spirit of wanderlust. When they emerge from the plane, the machine has transformed everything. The climate is different, the people often look and sound different, and our home (the hotel) has been radically reorganized. I'm convinced that children at this age have little idea what has really happened to them. Because the movement, gigantic in scope compared with anything else in their experience, was disconnected from their immediate sensory experience, and because they cannot yet understand space on the geographic scale, they apply the simplest explanation: the airplane has stayed in exactly the same place but has magically changed its surroundings. It's a bit like the famous Holodeck on *Star Trek*. Though we can smile and nod at their adorable confusion, the reality is that children and adults may not be as different as we think.

Children can be famously and sometimes hilariously confused about geography, especially in matters of scale, but we adults can be too. Especially when we are transported passively over long distances, it is easy for us to lose any real sense of the scope of space. It is a cliché to say that modern transportation has made the world into a smaller place, but this is exactly what has happened. Not only has the world become smaller but the spatial arrangement of features of the world has become less intelligible to us. This loss of intelligibility has come about in part because there is no longer a reliable connection between the effort that we expend to reach different destinations and their geographic relationship to us. For many journeys that involve flight, the time taken to travel from one's home to the airport can be greater than the time spent in the air. Intellectually, we are perfectly aware of the explanation for this— airplanes fly very quickly! But in terms of our implicit understanding of the spatial order of things, such distortions serve to loosen our shaky grip on the geometry of the world.

It might seem that this kind of mischief making with space, making the world seem smaller, should help us to overcome our inability to see the connectedness of different locations on the planet. But exactly the opposite takes place. Because rapid transit decreases the intelligibility of space, we throw up our hands in despair at making sensible spatial connections between things. Rapid transit has made it possible for us to be able to see much more of the planet than would otherwise have been the case—if I can scrape together the cash, I can go and witness firsthand the melting polar ice caps. But I question whether my ability to connect my home with the ice caps by means of an exotic air trip helps me to appreciate the absolute connectedness of the two places. Returning from my Arctic adventure, I don't arrive home with the feeling that the lines connecting me with the rest of the world are any more direct.

Given the way that our mind organizes and schematizes space, it is almost inevitable that there will be a harmful and dangerous schism between our inside and our outside worlds. If the parts of our brain that deal with space have strong preferences for enclosed views, and if we patch together a mental collage of space by combining these views, then any time a builder erects a wall, he is influencing our conception of the spaces in which we live. We can't avoid this by simply running through forests as naked nomads, basking in full-frontal contact with field and stream. But though complete integration with nature may be out of reach for most of us, it doesn't explain our modern tendency to run in the opposite direction, shunning natural settings for the air-conditioned comfort of our homes. Although our minds may be predisposed to detach us from real space, much more than a psychological predisposition has been at work in driving us from Eden to Gotham.

Jane Jacobs blames our tendency to insulate ourselves from nature on an impulse born of the European romantic movement, perhaps transported across the Atlantic in the guise of the New England transcendental movement espoused by Ralph Waldo Emerson and Henry David Thoreau.[2] At first, the connection between intellectual movements that cherished emotional contact with wild nature and the current difficulties in our relationship with the environment may be difficult to discern, but Jacobs's argument was that by raising wild places onto a pedestal, we convinced ourselves that life in our cities had nothing to do with the natural world. An impulse born of the noble desire to find truth in the forest had the result of increasingly polarizing urban and wild places. Whether through televised nature documentaries, urban zoos, or, if we can afford it, air-conditioned safaris through Ngorongoro Crater in Tanzania, we are taught to cherish nature, but from a distance. Perhaps in part because most of us fear true wilderness, we take nature in small, digestible gulps, contained in boxes, framed by the edges of a television screen, or even rendered in plastic facsimiles like lawn ornaments or fake houseplants. One might argue that in an urban environment this is the best we can do and that it is better than nothing, but underlying this attitude is still the general notion that nature has no place in the city and that only pristine wildernesses completely out of reach of any but the hardiest and most intrepid travelers really "count" as nature. In short, we love nature provided that it keeps to its place—out of our city streets and out of our homes.

Some architectural styles seem to almost shout the news of the divorce between city and nature. A good example is the on-again, off-again North American romance with the Georgian style, popular in Europe in the eighteenth and nineteenth centuries and revived during several periods on our continent. Georgian houses possess clean lines separating the enclosing walls of the house from the surrounding grounds, avoiding even the presence of a porch to draw a connection between house and yard. Most modern suburban homes appear to contain similar carefully structured divisions between the outdoors and the indoors. From the double or triple garage doors fronting the street to the imposing foyer that serves as a kind of hermetic airlock between the inner and the outer worlds, such homes make no pretense to be any part of a natural landscape. Even the expansive plots of land surrounding such homes tend to be bordered with high privacy fences, cutting off views and turning back yards into not much more than giant outdoor great rooms with groomed grass carpets and high blue ceilings. It isn't uncommon for the owners of such homes to spend most of their summer weekends in the sweaty business of ensuring that only those guests—plant or animal—that have been explicitly invited into the yard are able to remain, while all others are killed with chemicals or booted into the compost heap. If this is how we live our lives, then it isn't very surprising that we have difficulty making connections between our own actions in cities and the devastation of our environment, both urban and rural.

Our daily economic dealings with the world seldom have much to do with place, further breaking down our relationship with geographic space. A walk down the aisles of any grocery store will make this point abundantly clear. Where I live, the produce aisle may have some local fruits and vegetables for a few weeks out of the year, but most of the time the fare comes from farms that are thousands of kilometers away from my home. We joke about the fact that most of our household goods are made in Asia, a state of affairs brought about by a combination of the low cost of fuel to ship products and the low cost of labor to produce them, but this too tends to erase issues of place from our consciousness. In an era when any product or service can be obtained easily from anywhere on the planet, what reason do we have to think about the origins of things?

In 1996, William Rees, a bio-ecologist at the University of British Columbia, published an epoch-making book with Mathis Wackernagel, one of his graduate students. The main idea of *Our Ecological Footprint* is that it is possible for us to calculate a reasonably accurate average value describing the size of the area of land—our footprint—that each of us was using to take care of all of our earthly needs, such as production of food and goods and disposal of waste.[3]

The numbers that Rees and Wackernagel generated were shocking. For one thing, there were enormous disparities between the ecological footprints of members of typical Western countries like the United States and Canada and the ecological footprints of developing countries such as Bangladesh and Vietnam. In addition, if one calculated the global average for the human ecological footprint, it was clear that in the long term there was not enough land to go around. As if any other evidence were needed, Rees and Wackernagel's simple set of numbers suggested with stark immediacy that, as a species, our behavior was not sustainable. Either we would need to find a way to make our average footprint smaller or the population of human beings would have to decrease.

The idea of the ecological footprint has been enormously influential as a shorthand method for calculating the progress of a nation or a community on the path to sustainability, but what is most interesting about it in the context of our discussion of space is the great disparity between one's ecological footprint and one's geographical footprint. If we take the latter to mean the *location* of the mass of land that each of us requires in order to sustain our way of life, then for many modern humans, especially those of us who live in the superdeveloped West, the idea of a geographical footprint is almost nonsensical. As our goods, products, food, and services really come from everywhere, each of our individual footprints is globe sized. And if our lives are situated everywhere, then they are really situated nowhere.

Boosters of globalization have argued that the dissolution of the geographical footprint is a good thing. If different regions are able to specialize in those activities for which they are best equipped, then the overall quality of our lives will increase because redundancy is reduced. This argument can be used to justify a situation in which most manufacturing is situated in areas of the world like Asia, while other areas of the planet, like North America, have tried to specialize in knowledge industries. The difficulty, as Rees has said himself, is that without some form of regulation, the forces of globalization tend to reward those who are able to pay their workers the least and allow factories free rein over the environment, often with high profits but tragic human costs in both the short and the long term.

NATURE IS GOOD FOR US

One of the main arguments of this book is that our tendency to understand how the world is put together by knitting a series of visual glimpses into an extensive palimpsest of space has influenced many of the features of modern life. Rapid transportation, modern architectural forms, methods of building up cities, and the architecture of cyberspaces all reflect the ability of our mind to cope with the spatial fractures that are produced by these things. Then why should we become so hot and bothered by it? If we are lucky enough to have the kind of mind that can make sense of a world filled with the heavy distortions of space and time that have been wrought by modern technology, then why fight it? Why not put it to good use? After all, such technology has its advantages.

This is a powerful argument, and, though it might be an apostatic view for an environmentalist, I think it has some traction. We can't go back to being wild savages loping across the plains of the savannah. Instead, we need to find the way ahead. But in finding this way, we need to first make sure we understand where we have come from, why so many of us value our natural heritage, and what we stand to gain from its preservation. Leaving aside the apocalyptic visions of seas boiling dry from global warming or untold millions of human beings dying slowly from the cumulative effects of toxins in our water, soil, and air, there is a much simpler rationale for our wanting to find ways to heal the spatial rifts that lie between us and the rest of the natural world: contact with nature is good for our mind.

Like many who have made great contributions to our understanding of the natural world, Edward Wilson spent his early life mucking about in the woods looking for critters.[4] This informal childhood training positioned him well to make a lifelong habit of observing nature closely and eventually helped to propel him to a career as one of the most widely respected entomologists in the world. But what propelled Wilson onto the world's stage was less his painstaking documentation of the life histories of ants and more his bold claim that the tenets of evolutionary biology could explain many key aspects of human behavior—our social lives, our treatment of members of our family, and especially many aspects of our sexual behavior. Wilson, along with some early proponents of the new discipline that he called sociobiology, bravely argued that many human behaviors, ranging from homicide to altruism to adultery, could be explained at least in part by our genetic complement. In short, sociobiologists (these days more commonly called evolutionary psychologists) seek explanations for the ways in which our behavior might increase the number of our own offspring or the number of children that our relatives produce.

One of Wilson's more recent arguments, connected with his interest in the evolutionary origins of behavior but also springing from his concern about the accelerating rate of the planet's loss of biodiversity, is that human beings have a deep genetic connection and attraction to the natural world. This biophilia, as he calls it, transcends the pragmatics of finding ways to preserve enough air, water, and food to sustain life. It is a psychological, moral, and perhaps even a spiritual matter. Wilson suggests that our attachment to nature—the things that we like to look at, the kinds of landscapes that attract our eye, even the natural objects that we fear (snakes, for instance)—is etched into our genetic code. Early humans, he says, were served by this attachment to nature because it brought our forebears to select the kinds of habitats and objects that were conducive to survival.

It is not hard to think of evidence to support the biophilia hypothesis. More of us visit zoos each year than buy tickets to all sporting events combined. Perhaps this attraction to animals occurs because of an inherent understanding that natural areas of plenty will support a variety of animals. Those of us who are rich enough to live wherever we want command vistas at the tops of bluffs or hills, preferably with long horizons over water. These locations, containing plenty of Appleton's prospect and refuge, would also have been the most advantageous sites for early humans to see great distances without being surprised by predators. Human beings everywhere have a strong attraction to water. This makes perfect adaptive sense, as water not only provides protection from predators but also attracts other animals and plants that could be available as food.[5]

Even long before Wilson's biophilia hypothesis, substantial numbers of scientific studies had shown that we prefer natural settings over all others and that we prefer even sparse and scrawnylooking natural environments to almost all urban settings. More detailed investigations of the most attractive properties of landscapes have been carried out by showing people pictures of nature and asking them to rank them in order of preference. In virtually all cultures and from a very young age, there are universal preferences for the presence of water, prospect, refuge, complexity, and what has sometimes been called mystery—a more ephemeral property suggesting that exploration would quickly reveal some hidden features in the landscape. Some studies have even shown that we prefer tree shapes that would have most strongly predicted the presence of a thriving ecosystem in the savannah of our early forebears. Though none of these findings prove we are genetically presupposed to like natural settings that are good for us, the universality of these preferences across age, race, and upbringing are compelling in suggesting that we are all drawn to similar elements in nature.[6]

Mounting evidence suggests that exposure to natural settings can affect our mental and physical well-being. Surgical patients in rooms with windows looking onto natural features recover faster than those without such views. Views of landscapes or even the presence of indoor vegetation improves productivity in office workers. Nature scenes on the ceiling in dentist's offices lower stress levels and heighten pain tolerance. Psychological studies in which stress levels were measured directly suggest that we are mentally refreshed by a walk in the woods or even a few minutes of observing a John Constable landscape.[7]

In his popular book *Last Child in the Woods,* Richard Louv argues persuasively that our separation from the natural world behind the walls of our homes and offices is exacting a heavy human toll, especially for our children. Louv has coined the term "nature deficit disorder" to characterize the symptoms that accrue in children isolated from wildness, and he argues that such isolation can produce disorders resembling attention deficit disorder. More important, Louv suggests that finding ways to reimmerse our children in natural settings can actually cure such pathologies, making our children healthier and smarter.[8] If there is any merit to such arguments, then even if we don't believe that we are on the brink of an environmental apocalypse, we have ample reasons to be concerned about our lack of connection with nature. Even if one believed, as many in the previous generation seemed to, that technology would eventually find a way to make penance for our polluting sins, perhaps only by erecting giant bubbles filled with a chemical soup that could somehow sustain human life, this life would be a pale reflection of our present circumstances and perhaps not worth living.

How has this state of detachment come about? It is easy to point the finger of blame: at cheap energy (which has made it easy for us to distort space), at the romanticization of nature at the hands of poets (because their work, placing nature on a pedestal, has helped to polarize the differences between the urban and the natural), at the greed of those who can't understand that wealth is not always measured in dollars, and at our slowly evolving tendency to objectify the rest of nature perhaps set into motion by early scientific thinkers such as Descartes and Galileo. But though all these must have played a role in bringing us to our present plight, our unfortunate path was made possible only by the existence of a mind that was predisposed to see the world as a series of disconnected visual images, detached from one another and from their leafy anchors on the planet.

Early in my scientific training, I had a mentor who reminded me that there were limits to human understanding. What he meant was that simply by virtue of the way our brain is put together, certain facts of existence might be forever beyond our grasp, just as we are unable to detect the ultraviolet signatures of flowers that are so easily seen by bees. In the same way, I think it is possible that because of the organization of our mind and brain we can never experience the true oneness with nature craved by luminaries like Wilson and Louv. When we blink or move our eyes from one fixated object to another, our visual system shuts down for a moment so as to conceal the lurching, interrupted nature of the perceptual act from us, making the world we see hide its seams. We can no more completely understand the scope and shape of the space we inhabit than we can force our eyes to see during a blink. Not being able to make those connections, we can perhaps only dream of them.

Does this fact of physiology doom us to continue our spiraling course to a paved, sterile planet filled with dirty air, toxic water, and endless cycles of extinction? Not if, having understood the kind of creature that we are, we take what measures we can to bring true nature back into our lives, starting with an appreciation for our spatial connection to the planet. We may not be able to share the mute awe of a soaring hawk calibrating time and space as it closes in for a kill, but we can use our intellect, our wit, and perhaps even our penchant for technology to turn some of our weaknesses to strengths.

HOW WE CAN RECONNECT WITH SPACE

How do we begin to ratchet our soaring minds back down to a planet in peril? How do we re-anchor ourselves to the earth's surface in full knowledge that we have it within ourselves to take flight like Icarus on waxen wings and look down on our tired blue planet from the freedom afforded by the light air of high altitude? As with so many things, our best hopes for the future must be invested in the young.

We cannot turn our children into desert ants. No matter what tricks of technology or educational practices we bring to bear, there are finite limits on the accuracy with which we will be able to find our way through the world. However, we also know that children alive today, more than half of them born in intensified urban settings, have little knowledge of trees and birds, let alone the lengths and widths of wild places. Erik Jonsson, a retired engineer who has spent many years thinking and writing about human navigation, has pointed out that most formal psychological studies of navigating human beings have taken place in university laboratories using student volunteers. The typical profile of these volunteers is that of a man or woman in their late teens, from a fairly affluent and probably urban background.[9] One recent report found that fewer than half of students majoring in biology at a large Canadian university had even been on a camping trip.[10] Given this, Jonsson argues, we can only guess at the human potential for wayfinding. Methods of navigation used by traditional peoples, such as the Inuit eking out a living from the land or the intensively trained Puluwatese marine navigators, may differ markedly from those used by ants, bees, or homing pigeons, but there can be little argument that the ancient skills of these human wayfinders far exceed most of those used by modern urban human beings. Surely it is not a coincidence that such wayfinders also possess a remarkably strong bond with the natural world. As we've seen, geographical space, natural habitat, flora, and fauna are embedded deeply in the oral traditions, culture, and way of life of these remarkable groups.

Without parachuting our children into tracts of Arctic tundra or dropping them into oceangoing rafts, how do we begin to cultivate the same kinds of understanding of space and the same attachments to nature as seen in indigenous peoples? As Richard Louv points out, an important first step in getting our kids into natural settings to explore space is to get them out the front door. Once children are out of doors, they need places to explore, and this is where modern city planning principles fail them badly. In suburban settings, the greenspace that exists is most likely to be unattractive, barren, and filled with the noise of traffic. When green space allotments have been mandated by law for sprawling suburban developments, they have often been managed with a sad disregard for the facts of human psychology. Narrow strips of grass that meander behind suburban back yards may satisfy certain bylaws, but they do nothing to nourish our minds. These parched tongues of grass and little concrete playgrounds containing a few swings, a plastic slide, and perhaps one or two ride-on characters modeled after TV cartoon characters most often sit empty and unused. In urban centers, parks are typically small, rectilinear, and flat. Though children may be found in such spaces on occasion, the parks must be exceptionally well designed if they are not to become refuse dumps and havens for the homeless. The sad truth is that urban planning has taken too little heed of the needs of children, because those needs are not well understood. We don't know enough about how to win the battle to draw children away from television and computer screens and into the great outdoors. But what is certain is that such battles cannot be won simply by the artful design of spaces. In order to wean children from the strong enticements of the wired indoors, there must be a cooperative effort involving parents, city planners, and educators.

REDESIGNING FOR CHILDREN

The first step parents can take is to recognize that they themselves may be discouraging their children from outdoor play by providing supportive indoor environments that are altogether too alluring. Many children have TV sets in their bedrooms and unlimited access to computers that are connected to the Internet. In addition, our homes are filled with gaming consoles, Wiis, Xboxes, and Game Boys. Without wishing to reflexively dismiss the educational potential of these gadgets (we saw in the previous chapter that we can use such technologies to build and support our cognition of geographic space), their presence in the home can be stiff competition for the tiny grove of maple trees waiting at the end of the block. Parents provide these devices in part because their children ask for them and in part because parents recognize their educational potential. But they also do so to give children alternatives to outdoor play.

Suburban parents may perceive that the empty streets and parks of the outdoor world hold too many dangers, and those raising children in the city fear the risks posed by heavy traffic and strangers. For most of us, these fears are vastly overblown. Although population levels have risen dramatically, the overall numbers of child abductions, for instance, have not increased substantially in the past thirty years. In spite of such a soothing statistic, many parents have great difficulty in allowing their children to play out of sight, in the outdoors and away from the sheltering walls of the house. Those difficulties are compounded by dramatic media coverage of isolated events involving harm to our children, and it may even be that we are actively encouraged to harbor such fears by those with vested interests in keeping children seated before portals of 24/7 streaming commercial content issuing from all the screens in our house.

Are there ways for us to use the very technology that threatens our children's health and well-being to turn them out the door and into the wider world without raising our own levels of anxiety for their safety? I think so. As I write these words, I've sent my nine-year-old daughter on a walking trip to the grocery store. It's a little farther than I would normally let her stray from home, and it is in a part of town that, though comfortably filled with "eyes on the street," is frequented by a wide mix of different pathologies (including many neurotic parents like me). I've given her my cell phone. She calls when she arrives at the store and she calls when she leaves. If anything out of the ordinary takes place en route, she hits the speed dial.

Younger children, of course, can't use cell phones effectively, but enterprising technology companies have designed a variety of devices that can be used to track the location of our children using, for example, Global Positioning System technology. Bracelets can be configured so that they raise alarms when children wander beyond a certain range or when there is any attempt to remove the bracelet. In Denmark, the LegoLand theme park has partnered with Ekahau, an innovative technology provider, to design a "kidspotter" system in which children are fitted with bracelets that triangulate on position using the same types of signals that we employ to set up wireless Internet networks. By dialing a number on their own cell phones, parents can receive updates on the location of their children in the park, accurate to within a meter or two. There is little reason why such technologies could not be used to keep track of our kids in neighborhoods, parks, and other spaces that we might like them to explore.

Though such measures may strike some as smacking of Big Brother surveillance (one urban planner I discussed this with pointed out, correctly, that similar technology is used to track certain types of prison convicts or parolees), they can also be viewed as a way to subvert technologies that were designed to help us ignore or "work around" natural settings. These technologies can help us to ensure our children's safety while giving them the necessary autonomy to engage in the primary experiences with nature that they need to be healthy and well adapted as adults. Such tools may not be for everyone, and they are certainly unsuitable for older children, but for the segment of the population of parents who cannot stand the thought of their children wandering the great outdoors, they may provide some extra impetus that will get their nature-deprived kids across the threshold of the house and into some form of greenspace.

The possibilities for using geo-locating capacities of GPS systems to help our children engage with the natural world are particularly intriguing. I have begun to explore the world of geocaching with my 12-year-old daughter. Geocachers put a variety of interesting objects inside weather- and animal-proof containers, hide the containers outdoors, and then post the precise latitude and longitude coordinates on a website.[11] Other players visit the website, pick out the coordinates of caches that interest them, and then use a portable GPS to find the caches, make their own contributions, and record their finds in a visitor log at the cache site. Geocaching is a kind of scavenger hunt writ large in earth coordinates and, for older children, it offers a tempting combination of engagement with the outdoor world and the use of clever gadgets of technology. Geocaching has become a worldwide phenomenon, with caches being hidden near both the North and South Poles, throughout North America, Europe, and parts of Asia. As GPS technology continues to become less expensive and is more commonly included in other tools such as pocket computers and cell phones, there is tremendous potential for helping children of all ages to merge their natural inquisitiveness about gizmos with their innate biophilic impulses.

If the first line of attack is to find ways to get our kids outside (and to give ourselves permission to let them go out the door), the next challenge is even greater—finding ways to configure the environment so that they have opportunities for primary experiences with nature. Unlike mediated exposures to nature, such as field trips to local parks or tightly orchestrated school excursions to outdoor education centers, the most important natural experiences for children are those that they undertake on their own, spontaneously and without the benevolent controlling influence of parents or teachers. These natural experiences begin in the backyard tree, the local storm culvert, or the shady patch of weeds on the front lawn. Such locations provide the ideal backdrop for early ecological epiphanies that take place as very young children hold their first earthworm, watch urban squirrels leap from limb to limb, or notice the differences in smells of the leaves they crush between their fingers.

But though these early experiences are pivotal in helping to develop an awareness of and attachment to the natural world, they do not nurture children's awareness of space. To learn about space, children must move about, notice the connections between places, and challenge themselves to reconstruct larger views of space from brief snapshots. It isn't hard to imagine ways of redesigning the spaces between our houses to support this kind of wayfinding fun, but the ideal environments would be anathema to, for example, suburban developers intent on enticing customers by promising large, private building lots. One approach would be to replace the large back yards in such developments with much more modest spaces but to attach these to large, commonly held wild areas complete with complex groupings of native plants, bushes, and trees, interesting rock formations, and, wherever it can be done safely, water.

In his book on biophilic design, Stephen Kellert describes such an ideal setup in the development of Village Homes in Davis, California.[12] Houses face inward in small clusters of about eight homes each, providing the safety of enclosure to large naturalized areas held in common by the residents, who collectively plan and maintain shared common areas. Underground storm drains have been replaced by natural swales brimming with plant and animal life, and water from these swales is used to irrigate fruit trees. The homes themselves are well integrated into their natural context and built using sustainable practices, including north–south orientation to maximize passive solar energy and rooftop solar panels to supplement electricity use. Done well, such environments could not only give children opportunities to engage with wild places but could also be made spatially complex enough to support wayfinding challenges.

Even older, established neighborhoods can, with creative and open-minded participation from parents, be turned into enclaves of biophilic design. On my own urban street, a few blocks of houses ranging in age from about 40 to 100 years, some families have removed parts of the fences between their small back yards. Children are encouraged to play in both the front and the rear of the houses, and many games are explicitly designed to help them learn about the spatial connections between adjacent and remote back yards. One game that the children have dubbed man hunt, a variant on hide-and-seek, requires that all children but one hide and the "hunter" find each hider and lead them back to "jail." This game, mostly enjoyed in summer darkness while parents sit on front porches to socialize, encourages children to explore felt connections with spaces because visual cues are minimal. Such games, especially if embedded in a make-believe narrative, are so compelling for the children that all televisions and computers sit idle for hours at a time. Initiatives such as these can be relatively easy and free. The most important ingredient is a shifting of mental focus on the part of parents, a recognition of what is being lost by denying our children opportunities to explore space in nature and a willingness to replace the manicured but unused lawn with spaces of complexity, mystery, and challenge.

A similar initiative, though more ambitious, took place in Berkeley, California, when a lecturer in real estate slowly acquired an entire city block of residential properties, and removed fences, garages, and pavement to produce a large, interior shared greenspace. Early studies of the site showed that, compared to nearby areas with conventional lot layouts, residents were more connected with the outdoors and also with their neighbors.[13]

Schools, too, must play a stronger role in the development of spatial skills in children. The efforts that schools currently devote to generating novel methods to enhance literacy and numeracy should be matched by initiatives that encourage children to see beyond the walls of the classroom so that they feel connected with the larger world. Such efforts could begin with the design of the school itself. Most new schools that I've seen, built with efficiency and economy foremost in mind, consist of a series of low-slung rectangles set into large slabs of flat concrete playgrounds and surrounded by bare playing fields. Such designs do nothing to promote connections between outside and inside spaces. A teacher friend of mine told me recently that at one school where he worked, in the Canadian

Arctic, the community had embarked on a marvelous project to better integrate the physical setting of the school with the natural world. The organizers enlisted the help of some native elders, who immediately suggested that the classrooms and hallways should be filled with nature—plants, water, and large rocks—and that the building should be opened to natural light using skylights. Most of these measures were not carried out, falling victim to simple economics or perhaps some failing will, but there's no question that such changes in design, many of which should not be prohibitively expensive, would help children to make connections between the inside and outside worlds.

Most schoolyards are notoriously lacking in biophilic features. In part this is for the sake of safety—small numbers of teachers must oversee the outdoor activities of large numbers of children. But, as with the redesign of the natural spaces surrounding our homes, if there were ample recognition of the importance to children of cultivating a connection with natural spaces, we would be motivated to redesign schoolyards to satisfy both the need for safety and the need for engagement with wild spaces. In schoolyards planned with changes in elevation, natural groupings of vegetation, and perhaps even some explicit challenges to wayfinding such as mazes, children would be able to lose themselves, if only momentarily, and then in finding their way again obtain some sense of the connectedness of places. Not only would such environments be more conducive to free play but they could be used to structure planned activities that raise children's awareness of their own spatial strengths and weaknesses. Imagine a schoolyard game in which children name natural features, stitch them into stories like songlines, and then use these stories to plot routes. When such activities have been incorporated into the curriculum of innovative "survival training" camps for children, they have been remarkably effective. Children visiting wilderness camps years later still have vivid memories of places and routes because of the sticking power of these narratives. What has worked for the Inuit and the Australian Aborigines for thousands of years can also work for modern Western children if they receive proper instruction.

In recent years, as awareness of environmental issues has been brought to the forefront of academic curricula, schools have made some remarkable strides in teaching ecology. Many innovative programs have included outdoor classrooms where traditional pedagogic methods have been banished in favor of hands-on lessons that encourage children to think of the functional connectedness of natural ecological systems. Such programs are of tremendous value and should be supported by all parents. In sum what I am proposing is that the next step in healing the rift between our children and the larger world is to encourage them to see past the fractured views of space offered by their own nervous systems. This can be done by giving them opportunities to learn about the spatial linkages of things in environments that are specifically designed to facilitate such experiences.

GROWNUP FUN IN SPACE

As with so many things in life, in trying to put ourselves on a healthier trajectory with respect to our connections with natural spaces, our money and time are best spent on the requirements of children. But what about the rest of us? How can we make interesting spaces a larger part of our own lives? How can we break down our sharply drawn conceptions of a radically schematized space composed of a sequence of short glimpses collected while driving on freeways or hurrying along urban thoroughfares to complete the day's appointed tasks?

Many of the same principles that might apply to the redesign of play areas and schoolyards for children would have beneficial effects for adults as well. In cities, urban parks, instead of being calm oases existing apart from the main course of events and segregated from pedestrian traffic, could be placed as an integral feature of the urban environment. As much as possible, parks should present interesting wayfinding alternatives for pedestrians intending to walk from one place in the city to another. New York's Central Park, for instance, is tightly woven into the urban landscape in such a way that it not only serves as a landmark but is large enough to immerse visitors in wild areas. In addition, Central Park serves as the site of a vast number of cultural activities, ranging from theatrical performances to free rock concerts. Another successful greenspace on similar scale but with a very different design is Chicago's waterfront Lincoln Park, which is a recreational area but also integrates well with several adjacent urban neighborhoods.

City greenspaces are not always managed so well. The National Mall in Washington, D.C., for example, is enclosed by many of the country's most significant museums, yet it is nothing more than a vast rectangular space devoid of interesting features, natural or human made, or even much in the way of comfortable seating. This huge piece of urban greenspace holds vast potential but is currently wasted. Similar spaces, though much smaller, can be found in midsized cities throughout the world. Open rectangles with paved pathways, brick planters, and perhaps one or two trees may satisfy city plans that mandate greenspace density, but they do little or nothing to place us into contact with natural settings.

On my own walk to work in Waterloo, a typical midsized North American city, I pass through a small, dense urban core and then enter a large city park with forests, fields, winding trails, and even a small area housing a number of interesting animals, both local and exotic. The park connects the urban core to another belt of development, which contains university buildings, student housing, and some commercial areas. There are countless ways of passing through the park to get to my workplace, many of which carry me into small groves of trees surrounding a creek. In these groves, I see otters, squirrels, chipmunks, waterfowl, and a profusion of songbirds. I am lucky to live in a city that has had the foresight to preserve slightly more than 40 hectares of greenspace close to the urban core and has, so far, resisted the wiles of developers, but many of the best features of this park could be managed in a much smaller space.

Another nearby preserved area consists of a small ravine filled with dense forests of birch and maple trees crisscrossed by trails and punctuated by a sinuous creek that serves to carry storm runoff away. I have spent many years walking and running in this ravine of less than eight hectares, and it seems as though every visit brings a new view or an unexpected connection to light. The undulating trails make my impression of my precise location just ambiguous enough to bring me to a pleasant state of surrender, curiosity, and attention. Unlike the larger park, this area makes weaker connections with the rest of the city. Many who live in Waterloo are not familiar with the small woods. Placed in a different situation with respect to the surrounding area, woven into the core of the city, such a space could not only serve as a cool respite from the busy grid of streets but could also challenge our spatial senses, remind us of the fragility of our grasp on place, and encourage us to pay closer attention to our ephemeral connections with the rest of the planet.

Even in suburban areas some fairly simple measures could effect profound changes in the ways that the landscape stimulates human awareness of space. I have already mentioned the possibility of corralling shared wild space by removing fences between overly generous back yards to produce larger, wilder spaces. As well, the generous setback of houses from the street leaves vast amounts of space with the potential for redesign. Roads could be narrowed, which would increase the amount of available space considerably and possibly redirect traffic entirely in some areas, leaving the suburban equivalent of an urban pedestrian boulevard.

All these new measures encouraging pedestrian life, of course, will work only if they are supported by mixed use. People will walk only if there are places to walk *to*. Though current bylaws might preclude such mixed uses in many areas, the day will come when the rising cost of automobile travel will make mixed use in suburbs, incorporating businesses, market gardens, community halls, and other public spaces, seem like a much more attractive proposition. When the suburbs contain favorable destinations, they can be connected to one another through interesting, challenging, but intelligible networks of pathways.

ENGAGING ALL THE SENSES

An entirely different approach to discouraging our tendency to chop up space into a sequence of discrete views is to enlist the participation of our other senses. Japanese gardens are arranged in such a way as to take command of all the senses rather than just recruiting our gaze. Though there may be many beautiful sights in such a garden, what possesses us is the way in which our visual experiences cohere with sounds (of running or falling water, for instance) and with the sensations produced by our own movements as we follow meandering paths through the garden. Our eyes follow our feet, our ears, and even our hands as we are compelled to imagine the tactile properties of the smooth stones whose careful placement in the scene enhances its overall polysensory effect. It might be unreasonable to propose that entire city blocks could be designed to replicate the sensory absorption produced by a small Japanese garden, but some of the same principles could be used.

Architects who are designing concert halls are naturally intensely interested in the aural environment of the interior of the hall, but other kinds of buildings and even streetscapes possess distinctive auditory properties. Consider the difference between entering a cozy, carpeted space in an intimate den and walking through the cavernous lobby of a large bank building. Even with your eyes closed, you would notice a stark difference in the auditory properties of the spaces and the manner in which the spaces either absorbed or reflected the sounds that you produced as you moved through them.

There are some interesting parallels between the ways that our own movements shape the aural environment that surrounds us and the mechanisms that are used by strongly path-integrating animals like ants. In both case, space is being marked out by our own actions in a way that isn't true for vision. For one thing, many of the sounds that reach us, especially as we move about in buildings, are reflections of self-produced sounds such as footsteps. As Barry Blesser and Linda-Ruth Salter describe it in their book on aural architecture, *Spaces Speak, Are You Listening?* a critical difference between the sights and sounds of architecture is that we don't usually make our own light, but we do make our own sounds.[14]

In natural settings, we experience a closer link between the senses than is true for most built settings. As we walk through a forest, the sounds of our own footsteps meld with other sounds such as those produced by birds and insects, the sounds of wind in trees and other plants, and the sounds of moving water. In nature, there are fewer surprising transient sounds equivalent to the urban honking of a horn or the assault of jackhammer on pavement. Instead, an integral natural logic to the sensory landscape encourages us to distribute our attention more widely over space so that we become much more sensitive to subtle events.

I spent a portion of the writing time for this book living in a very small fishing village on the east coast of Canada, and I fell into the habit of taking long daily walks along deserted stretches of road, trail, or beach. Like so many others before me, I discovered that these walks exerted a powerful positive influence on my patterns of thought both while en route and for many hours afterward. In the beginning, I thought that much of this therapeutic effect had a simple physiological cause—the movement simply helped to pump more freshly oxygenated blood to my brain. But I came to realize that an important component of these walks was the manner in which sensory experiences coming to my eyes, ears, skin, and nose were perfectly in accord with the rhythms of nature. The lapping waves, the crunch of stones or dry leaves underfoot, and the smell of salt and pine combined to produce an ineffable sense of place that became so strong that, like an Aborigine wandering the outback, I began to associate particular ideas with locations. From time to time, I would even revisit locations now resonant with my thoughts in order to clarify a concept or hash out a difficult ambiguity. The key element of these personal songlines, I came to realize, was that the perfect fit between what I could see, hear, feel, and smell made me recognize more profoundly than perhaps at any other time in my life that not only was I in a real *place* but my presence, move-ments, and the content of my mind helped to define that place. My thoughts and sensations strapped me to the earth as securely as a scurrying ant carries a homing beacon for its nest.

My own experience was an extremely lucky and unusual one. Few of us have the chance to sequester ourselves away from the busy chaos of life for several hours every day to be alone with our thoughts and with nature. But is there anything that can be learned from such experiences, and adapted to the pace of modern urban life, to encourage us to reconnect with the multisensory dimensions of real space and time? Inside buildings, it is certainly possible to do a great deal to sculpt the aural environment, by varying the texture and shape of walls, changing the ceiling height, and perhaps even adding water features. Though it has not been attempted as far as I am aware, it should also be possible to generate aural experiences that are specifically geared to the movements of pedestrians. This might be a particularly effective strategy to use in long passageways, such as the subterranean connections between office buildings in urban centers.

Given a city's large open spaces, public access, and complex mix of uses, deliberate attempts to construct soundscapes there are bound to be much more complicated than such efforts inside buildings. Some very creative approaches to this challenge have been adopted by cutting-edge artists interested in urban soundscapes. Andrea Polli's *NYSoundmap* was an installation at the 2006 Conflux Festival, a meeting devoted to psychogeography held every year in Brooklyn.[15] The installation included a clickable webmap that could be used to navigate the aural environment of the streets of Brooklyn. In addition, Polli provided instructions to those interested in heightening their attention to urban sounds while they were walking. At the same meeting, Sawako Kato, a sound sculptor who straddles venues in Tokyo and New York, unveiled a work entitled *2.4Ghz Scape,* in which sound-processing technology was used to convert ambient signals from Wi-Fi sources into a soundscape that accompanied a walker through any location where such signals could be detected. As Wi-Fi can now be found in almost any urban center (I must still say "almost" as, when I recently asked a server at a Florida bar whether Wi-Fi was available, she consulted the bartender, who looked for the appropriate bottle on the shelves behind him), the possibilities for taking advantage of these signals to produce soundscapes that might accompany a walker wearing a portable listening device are at least interesting.

In a variant of this idea, Mark Sheppard, a professor of architecture at SUNY Buffalo, has developed the Tactical Sound Garden.

Users are able to plant or prune individualized sound signatures in particular locations in the city that are defined by complicated combinations of public Wi-Fi signals. Freely available software can then be installed on a cell phone or pocket computer so that a user, wandering through the city, is able to listen to the garden as they walk the space. Interestingly, Sheppard uses an explicit community gardening metaphor to describe the project, likening the shared manipulation of sounds to the nurturing of a garden.

Though devices such as these might be adopted only by the technological cognoscenti at first, it would be entirely possible to adapt a wide variety of devices ranging from smart phones, Black-Berrys, or even MP3 players to provide place-specific aural content to accompany urban pedestrians. Electronic sound gardens are a far cry from the sensory experience of a country walk, but it would not be impossible for such technological devices to become an acceptable way of generating multisensory urban experiences that could help fix us to a place. Further, if these sound gardens could be specifically geared to reflect an important feature of our natural environment, they might play some role in helping us to feel our connections to even larger spaces.

Andrea Polli's project *Airlight Taipei* may be a step in this direction. This installation provides a repeating audible signal that reports the measurements of air pollutants by a local monitoring station. One could imagine more complex auditory signatures to signal relationships between local activities and their effects in the larger environment.

Educating children to make better connections with space, redesigning parks to make them functional parts of the urban core, and tapping into Wi-Fi networks to let walkers hear sounds that correspond to their own movements through city streets might seem like paltry suggestions in the face of a world that many experts tell us is at risk of drying out and catching fire within as few as two generations. My suggestions for redesigning cities, suburbs, and parks might risk comparison with a call to rearrange the deck chairs on the *Titanic*. There would be some merit to this criticism if I were suggesting that adopting the measures I have described in this chapter could reverse our current trend toward increasing destruction of our natural heritage. I am not so foolish as to think that my suggestions will be of any benefit unless many other, more dramatic measures are put in place. If we are to save ourselves and the plants and animals that we live among, we will need to rapidly develop alternative sources of energy. We will need to massively invest in carbon-sequestration technologies that capture the carbon produced by certain types of industries (oil sands extraction, for instance) and return it to the ground, and we will need a fast rollout of many other types of technology to halt and then reverse the trend that is beginning to boil our waters and scorch the earth. These measures will take time to put in place, and there is not a moment to be wasted.

But as much as I am aware of the urgency of our plight, it also strikes me that we have known for decades about many of the problems that are now acute. One might argue that only a simpleton could think we could get away with pouring poisons into the air and water of a finite system forever without eventual catastrophe, yet this is how we have behaved. What is even more striking is that we have long understood clearly where much of the excess greenhouse gases are coming from (I remember learning about this as a student in elementary school), and it is within our means to begin to halt the process. Inflated or nonsensical geographic footprints, unflinching appetites for vast quantities of cheap manufactured goods, and a feeling of entitlement to any goods from any location on the planet provided we can scratch up the cash to pay for them all contribute to our environmental crisis. Large coal plants in the United States belch out pollution because people demand and expect uninterrupted services and ice-cold air-conditioned environments in huge houses filled with all manner of power-guzzling appliances on days hot enough to cause the grass to literally catch fire. We burn enormous quantities of gasoline in a daily commute from jobs in the city to houses on the outer fringes of the suburbs, where we convince ourselves that we are in contact with nature, and then we spend our weekends using machines run by inefficient gas motors to shape, chop, grind, pulp, and blow away any parts of *real* nature that have the audacity to invade our high-fenced outdoor great rooms.

These are not technological problems but psychological ones. The real reason that our planet is dying is not the coal plant in Ohio, the auto manufacturer in Detroit, or the mammoth Asian factories spewing poisoned water into rivers. These are merely the symptoms. The late Stephen Jay Gould, an evolutionary biologist at Harvard University and a masterful popularizer of the history, lore, and wonder of nature and life, argued famously that "we will not fight to save what we do not love."[16] To this argument, whose deep truth I cannot doubt, I would only add that we will not love what we cannot see, hear, touch, taste, and smell. The measures I have described in this chapter are not meant as a cure-all but merely as starting points on the road that must be traveled if we are to find ways to remember that the spaces of our perceptions and thoughts must connect with the spaces of our bodies and the fields, forests, streams, and oceans that lie beside them.

CHAPTER 12
THE FUTURE OF SPACE

Our children will pay for our joyride.
ROBERT KENNEDY JR.

Our journey has been a long one. We began in the sediments at the bottom of cold mountain lakes where single-celled animals navigate using onboard iron compasses locked on to the earth's magnetic core. We clambered through thickets of different types of animals, each offering new and clever solutions to the overarching question of every animal's life: where am I? Along the way we discovered things about our own unique way of drifting through space using a compass course set slightly off kilter from the main path. Like Icarus soaring high above the planet on wings constructed by his father, Daedalus, the first engineer, we have minds designed to look down on a space of our own conception. But like those who died in the legendary labyrinth of Knossos, Daedalus's other famous construction, we abandon our connections to earth too quickly, lost unless we cling tenaciously to the artifice of Ariadne's thread. All who have mastered survival in wild places, be they ancient seafarer, wandering Inuit hunter, or modern woodsman, have understood in their bones this delicate balancing act between our soaring mental spaces and our feet planted precariously on solid ground.

We have much in common with every animal from the single-celled amoeba to the bear lumbering confidently through the boreal forest. All animals that move must be able to find the things they need and avoid those that can harm them. At the very foundation, this means possessing the hardware to systematically reduce or increase our distance from identifiable targets, ranging from the box of cereal on the supermarket shelf to the fanged predator lurking in the bushes. Though some clever tricks might be required to coordinate different parts of our body (eyes, feet, hands) with the things that we see, such target-orienting tasks hardly require sophisticated mental maps of space.

When we need to find our way to invisible or remote targets using more complex routes, more sophisticated means of wayfinding must come into play. Like most other animals, we have found ways to use landmarks both to mark positions and to point the way ahead. Just as wasps and birds may use visible features of the terrain to mark a nest or a cache of food, we can learn to use both obvious urban landmarks like the Statue of Liberty or, with training, more subtle features of the natural environment such as a particular species of tree or even a notable formation of rocks to send us on our way. Many early human cultures much more attuned to the land than we are learned to embed such subtle features into long narratives that helped them to locate themselves but also helped to make the emotional and spiritual attachments to place that seem to be lacking in our modern way of life.

Some of nature's most sophisticated navigators are capable of constructing accurate maps out of brain tissue, letting precisely organized connections between neurons stand in place of the grid lines on the paper maps that are more familiar to human beings. Though people also make maps with minds, they are most often composed of a slippery, rubbery substance where distance and angle mean little but connections between one thing and another count for much. This type of map, more flexible and less constrained than the metric map offered for sale at a gas station, can change in size and shape according to need, purpose, and even mood. Such maps, though they might not lead a bee to nectar, can serve us well by making it easy for us to remember simple, well-traveled routes and to communicate these routes to other people. But in the face of the least uncertainty, unfamiliarity, or unexpected change of course, such maps can leave us quickly and irredeemably lost. Not only this, but such topological maps, based largely on collections of quick glimpses, snapshots, and vistas, can generate idiosyncratic views of how larger spaces fit together. These strange views, in turn, can have strong influences on how we think about and behave in our homes, the places we work, our cities, and our greenspaces.

Now, in the twenty-first century, we have been successful in using our ingenuity, along with vast amounts of energy, to essentially eradicate real physical space from much of our lives. Most residents of modern cities live in climatically controlled environments, carefully shielded from the outside world in every sense, but particularly in the visual one. When we are outside of our dwellings, our views and vistas consist of the square corners of the carpenter and flat ribbons of road and sidewalk. The most salient connections that we make between one place and another are often forged by electronic switching stations and fiber-optic cables. Though this may not always be an aesthetically pleasing way to get through a day, modern life has its advantages. Acting as though geography does not exist allows many of us to eat and drink whatever we want, do pretty much whatever we feel like, and talk to anyone who is within reach of a telephone, a cellular network, or a computer. More than at any time in our past, human interactions are unfettered by the laws of physics. As I have tried to argue in this book, the makeup of our brainware has not been responsible for our current state, but a brain that allows us to weld together collages of space from one long series of snapshots after another has helped to lay the groundwork for such a transformation. What reaches a kind of apotheosis in the highest accomplishments of virtual reality technology is also apparent in much more humble contexts. We human beings don't just live in space—we make it. By letting go of Ariadne's thread, we may lose our grip on the planet's surface in a way that would be anathema to an ant, a bee, or even a desert scout in a preliterate human society, but we can gain immeasurable riches when we take mental flight, looking back down on geography of our own devising.

What happens next?

CAN SPACE END?

It might seem hard to imagine any kind of impediment or braking force on the human tendency to effectively shrink all space to a single point. As foreseen by Paul Virilio, the notion that we might not need to frame our existence in anything like Euclidean geometry suggests that the current trends to instantaneous communication, virtual existence, and a kind of distributed embodiment in which we live everywhere and nowhere should only accelerate. As technological developments continue, especially those in the realm of computing and information processing, the most straightforward prediction is that we will continue to leverage our current abilities to produce artificial immersive experiences so that physical movement from place to place becomes less and less important. If we can raise a convincing simulacrum of a beach in Tahiti from our basement VR theater, then why bother climbing into our flying cars to go to the real thing? If we can flick a switch to embody a fully equipped avatar sitting at a conference table on the other side of the globe, then why put up with the cramped quarters of an aircraft cabin?

This conjures images of a kind of post-human existence in which we would live in ways functionally equivalent to brains in jars jacked into computer terminals, and there are reasons to be skeptical that this is how our future will unfold. Technology has made it possible for us to warp and weave our way through space in paths that defy physics, gravity, and essentially everything other than the speed of light, but there are already plenty of signs that there are limits to the extent to which we can allow such technologies to rule over our spaces while maintaining our happiness, or perhaps even our sanity. Although we may have left some of the wayfinding skills we share with other animals in ancient dust, we still sometimes seem to respond to the call of old biological circuits that remind us of the importance of place.

The current fascination with GPS and geo-coded data is one sign of an inner sense of the importance of the *wheres* of our lives. Michael Jones, the chief technical officer of Google and the developer of geospatial applications such as Google Earth, describes a romantic fascination with place-based computer data that draws hundreds of millions of users to such applications first to play and explore but ultimately to learn and to draw connections between themselves and the real places of the world.[1] The best example of this is the Crisis in Darfur project, a joint effort between Google and the United States Holocaust Museum, in which virtual visitors can zoom over villages burned to the ground by Sudanese soldiers.[2] Viewers can interact with villagers one on one by flying in close to see and hear firsthand accounts of atrocities from victims and their families. As always, a part of what moves us is the narrative, but being able to attach the story to places that can be seen on the screen adds considerably to the impact of the presentation and leads to a greater likelihood of action on the part of remote bystanders sitting at computers in another hemisphere. As has ever been the case, attaching story to place enhances the power of both.

The myth of the labyrinth speaks to a deep human need to maintain a grip on our own implacement in the world—how we are connected to a specific location on the planet. A story in which the central setting is a dark set of unintelligible caves that can be navigated only by maintaining a tight hold on the slender thread offered by Ariadne resonates with the primal fears we attach to the state of being lost. More innocuous forms of labyrinths, happily free of voracious Minotaurs, have been used for hundreds of years to help people maintain a sense of connectedness to location while moving through space. One of the grandest examples of such a labyrinth can be found in the floor of the Cathedral of Our Lady of Chartres, near Paris. Befitting the Gothic symbolism of the whole cathedral, the dimensions of the labyrinth laid into the floor are replete with sacred geometry and numbers, but the floor has a greater purpose: visitors complete a symbolic pilgrimage by walking the circuitous path to the center. In some ways resembling a Japanese garden that is designed to connect movement with vista in a carefully orchestrated experience, the labyrinth reminds us of our connections to the ground and invites us to examine explicitly the manner in which our own movements influence our relationships to places.[3] In recent years, the popularity of labyrinths has increased notably. Patterned floors for use as labyrinths are being installed in community halls, in church basements, and on private land. Though the ostensible aim of such projects is to offer people a spiritual tool that enhances meditation while walking, such constructions also serve to renew and celebrate the human connection to place.

Place-based political movements have become an increasingly important force on the world scene. One of the most prominent modern instances of political reactions to the homogenization of space occurred in Chiapas, Mexico, when poor sustenance farmers reacted violently to economic threats to their existence caused by the passage of the North American Free Trade Agreement. The so-called Zapatista Revolution galvanized growing levels of dissent in the face of sweeping forces of globalization that were themselves being generated by our ability to communicate across global distances in an instant and move goods anywhere cheaply using abundant and inexpensive sources of energy. What is most often at the root of such place-based protests is an objection to the manner in which transnational corporations are able to submerge local economies and produce very real human suffering, but these uprisings can also be seen as expressions of unhappiness with a world that is shrinking, dropping its borders, and pushing *where* issues into irrelevance.[4]

Other, more gentle place-based initiatives involve renewed understanding of the connections between our lives and the places we inhabit. The popularity of the 100-Mile Diet, in which consumers are encouraged to pay close attention to the geographic origins of the food that graces their dinner table, is but one example of a generalized encouragement to "think local" whenever we are able to.[5] The arguments in favor of such "localism" usually have more to do with the pragmatics of wasted energy, pollution, and bad farming practices than they do with the importance of human attachment to place. But this is not to say that forming such attachments is not an important part of the appeal of such approaches. Peter Mayle's *A Year in Provence*,[6] for example, may seem to be a romantic account of a bygone way of living, but it may be more like salve on psychic wounds caused by the fracture between the disconnected spaces of the modern world and our tightly knitted connections to places that our genes remember from ancient times.

Collectively, these signs of resurgence in our understanding of the importance of the *where* element of our lives suggest that even if human civilization found a way to continue to exist and to follow our current technological trajectory for a good deal longer, we could not be persuaded to abandon our devotion to physical place, no matter how seductive the realms of the virtual might become. Even if a virtual world looked and sounded exactly like the real thing, there is every chance that simply knowing it wasn't real would be sufficient grounds for us to reject it. Though our mind may in some ways condemn us to live in a nether world of semi-detachment from space, we will continue to recall enough of this missing connection to yearn for it and so include it as an important element in our lives.

PICKING UP THE PIECES OF SPACE

The modern pace of change in every realm from the political and social to the scientific and environmental is so rapid that trying to imagine the future is almost impossible. What seems safest to say is that any "business as usual" vision of our future is certainly delusional. Cheap energy will soon dwindle, and conflicts over resources will heat up as competition arises for the remaining pools of oil, water, and food. Some can muster no optimism that we will find any rational way to accommodate such rapidly changing prospects and that ultimately, when the lights go out and the computers shut down, having lost the tools and knowledge we need in order to do without, we will essentially be back in the Stone Age.

At the same time as the sun is setting on the era of cheap energy, our environmental problems heat up. Sea levels are rising, ecologies are changing faster than we can monitor them, and the rate of extinctions continues as an unabated trend that will decimate the planet's biodiversity. Unless such trends are slowed, halted, and eventually reversed, we can expect to see famine, drought, super-storms, and suffering on a vast scale.

In the face of such daunting prospects, it can be difficult to face each day, let alone to look into one's children's eyes and offer reassurances, other than to tell them that each of us is doing what we can to assure that the planet and our species will have a future. Yet for all the doomsaying, nobody can tell us with any certainty that the time for hope has passed, and provided this is so, there is still time for action.

I have argued in this book that the way our mind parses space is one of the root conditions underlying our astonishing ability to accomplish such technological wonders but also to neglect our stewardship of our planetary home to the extent that we risk losing it. Along the way, I have made recommendations for positive changes to help us work against our proclivity to carve up large-scale space into smaller, disconnected fragments. Some of these recommendations have to do with ways to reorganize our lived spaces, and others with ways to use space-collapsing communication technologies to forge connections with the remote quarters of the planet that are affected by our very local actions. But I have also taken pains to point out that such suggestions will not by themselves reverse our current course.

Massive technological intervention in the form of carbon sequestration and the development of alternative fuel sources will also be required if we are to salvage any semblance of our current way of life, but even these happy developments will allow us to dodge the dual bullets of climate change and energy shortages only if we also find ways to effect huge changes in how we think about our lives, and if we reorder our values and make reality checks on our wants and our needs. In the absence of a "business as usual" model, the alternative will be to work much harder than we do at present to live locally, joining small social networks where we can share expertise and resources, avoid unnecessary travel, and minimize our ecological footprints. Another book could be written to flesh out these simple ideas (and many already have been), but regardless of what particular recommendations are eventually offered, key to our future existence and happiness will be a thorough understanding of how our mental connections to space influence our understanding of place. By trying to make explicit the links between how our mind comprehends space and many aspects of our modern lives, my goal is to contribute to greater awareness of the manner in which our neurological hardware, the heritage of our biology, must be accommodated if we are to find a place in the cosmos that is comfortable, happy, and sustainable in the long term.

Where are we? We are *here* now. Our future together depends on finding ways to understand and to feel the deep truth of that connection between person and place.

ACKNOWLEDGMENTS

I thank the people who helped make it possible for me to have a career in science, including John Yeomans, Mel Goodale, and Barrie Frost, and colleagues who believed in me and encouraged me to write, including Mike Dixon, Ken Coates, Pat Wainwright, Barbara Bulman-Fleming, Derek Besner, and Phil Bryden. I'm also grateful to those who read, commented on, and discussed parts of this book—or simply helped to keep the intellectual fires burning by fueling me with amazing ideas—including Hong-Jin Sun, Jack Loomis, Jim Blascovich, Mark Zanna, Marcel O'Gorman, and Cordula Mora. I give special thanks to my colleagues at the Waterloo School of Architecture, especially Thomas Seebohm (whom I miss every day—we were just getting started), Philip Beesley, and Robert Jan van Pelt. These individuals not only made some of the experiments described here possible, but they also helped me understand what an old-time experimental psychologist might be able to contribute to designing the built world.

Some great students have helped me with experiments and spent time with me discussing some of the ideas in this book—in particular, Deltcho Valtchanov, Justin Perdue, Kevin Barton, Brian Garrison, Punya Singh, Ela Malkovsky, and Leanne Quigley.

I'm lucky to have friends who listened to my complaints about this project, shared the joys of it (and an occasional bottle of Scotch), and offered suggestions both helpful and deliberately ridiculous. This circle includes Richard Akerman, who sent me new and interesting snippets at least once a day; Brandy and Ian Cameron, who shared the Leverette Transform; Janet and Chris Piers, who showed me the Arctic; Peter and Anita Mason, who took me where the bears were; Andrew Brooks; Kate and Brooke Oland; Mark Wright; and Marysia Bucholc. Janeen Armstrong, whom I've met only in cyberspace, flipped me a great title, which helped bring about the book's cover design.

I've had a ball learning about the trade publishing world from Jim Gifford and Nita Pronovost at HarperCollins Canada, and Dan Feder and Melissa Danaczko at Doubleday. Without these people, my book would have been long and boring. I'm grateful every day for "Special Agent Bukowski," my literary agent, Denise, who worked very hard to make this book happen, pushed me to do my best work, and saved me from certain doom on many occasions.

Finally, I thank my family. My brothers, Norman and Martin, and my sister, Jennifer, have been in on this idea from the beginning. My children, Sarah, Emily, Jessica, Rebecca, Mei Ling, and MacLaughlin, followed me on a wild ride, including a year spent pacing a Nova Scotia beach (where much of the writing was done) with Gilbert the scruffy wonder dog. My wife, Karen, has been my mentor, advisor, accountant, confessor, rock, soul, life coach, and fan. I'd get lost with her anywhere.

NOTES

INTRODUCTION: LOST AND FOUND

1 Lynn Rogers, a field researcher with the U.S. Department of Agriculture, describes some of the navigational feats of black bears in her article "Navigation by Black Bears," published in the *Journal of Mammology* 68, no. 1 (1987), 185–188.

2 Edward Casey's extraordinary books document some of the interplay between philosophy, both ancient and modern, and how we use and think about space. His book *Getting Back into Place*, published by Indiana University Press (Bloomington) in 1993, is his most accessible work.

3 Briane Greene has written a fascinating and mind-boggling account of space and time as the physicist sees it entitled *The Fabric of the Cosmos: Space, Time and the Texture of Reality* (Knopf: New York, 2004).

CHAPTER 1: LOOKING FOR TARGETS

1 The epochal paper by McKay and colleagues was published in the high-profile journal Science (D. S. McKay et al., "Search for Past Life on Mars: Possible Relic Biogenic Activity in Martian Meteorite ALH 84001," Science 273 [1996], 924–930).

2 A group showing evidence that magnetite similar to that found in the Martian asteroid could be formed by simple non-biological processes was led by David McKay's brother Gordon. See D. C. Golden et al., "A Simple Inorganic Process for Formation of Carbonates, Magnetite, and Sulfides in Martian Meteorite ALH84001," *American Mineralogist* 86, no. 3 (2001), 370–375.

3 Tom Collett and Lindsay Harkness wrote an entertaining chapter on the variety of solutions to problems of distance vision entitled "Distance Vision in

Animals," in *Analysis of Visual Behavior,* edited by D. J. Ingle, M. A. Goodale, and R. J. W. Mansfield (MIT Press: Cambridge, 1982), pp. 111–176.

4 Valentino Braitenberg's marvelous book *Vehicles: Experiments in Synthetic Psychology* (MIT Press: Cambridge, 1984) describes simple thought experiments showing how seemingly complex behavior can emerge from simple combinations of wheels, motors, and sensors.

5 Yarbus's early experiments on eye movements are best described in his book *Eye Movements and Vision,* first published in Russian in 1965 but translated into English by Basil Haigh in 1967 (Plenum Press: New York).

6 James Gibson continues to influence a vast number of scholars in fields as varied as architecture, fine art, psychology, and urban planning. Specialists in each field claim to understand what he meant to say, yet each group takes something quite different from his words. In part, this is probably because of the opacity of some of his language. Like others trying to express entirely new ideas, Gibson had to strain language to contain his thoughts. His first extended discussion of optic flow can be found in his book *The Perception of the Visual World* (Greenwood: Westport, CT, 1974). Edward Reed has written the major biography of Gibson (*James Gibson and the Psychology of Perception,* Yale University Press: New Haven, CT, 1989), which contains much interesting detail about his stint in the military and its influence on his thought.

CHAPTER 2: LOOKING FOR LANDMARKS

1 Some details of Tinbergen's remarkable life can be found in a brief autobiography that he wrote on the occasion of his Nobel Prize in 1973. The text can be found at http://nobelprize.org/nobel_prizes/medicine/laureates/1973/tinbergen-autobio.html. Hans Kruuk has also written a nice biography of Tinbergen entitled *Niko's Nature: The Life of Niko Tinbergen and His Science of Animal Behaviour* (Oxford University Press: Oxford, 2003).

2 Tinbergen's classic digger wasp studies are described in his first book, *The Study of Instinct* (Oxford University Press: Oxford, 1951).

3 It is difficult to think of a single good source that summarizes Tom Collett's extraordinary contributions to our understanding of how insects navigate. His work, always published in top-quality journals, commands a great deal of attention among biologists and has made great contributions to the field. Two recent technical papers covering some of the ground I mention are T. S. Collett and M. Collett, "Memory Use in Insect Visual Navigation," *Nature Reviews Neuroscience* 3 (2002), 542–552, and V. Durier, P. Graham,

and T. S. Collett, "Snapshot Memories and Landmark Guidance in Wood Ants," *Current Biology* 13 (2003), 1614–1618.

4 The experiment conducted on Inuit and Tenne was published as J. W. Berry, "Tenne and Eskimo Perceptual Skills," *International Journal of Psychology* 1 (1996), 207–229.

5 A discussion of spatial modifiers in Inuktitut can be found in Judith Klein-feld's paper "Visual Memory in Village Eskimo and Urban Caucasian Children," Arctic 24, no. 2 (1971), 132–138. The example sentence comes from R. Gagné's paper "Spatial Concepts in the Eskimo Language," in *Eskimo of the Canadian Arctic*, edited by V. F. Valentine and F. G. Valee (McClelland and Stewart: Toronto, 1971).

6 Cora Angier Sowa has written a fascinating account of the World Trade Center as a "mythic place" in which she tries to account for the human reaction to its loss. Her essay is available at www.minervaclassics.com/wtcholy.htm.

7 Direct comparisons of the use of landmark configuration in animals and humans have been carried out in a series of experiments spearheaded by Marcia Spetch at the University of Alberta. One good source is a paper written by Spetch et al., "Use of Landmark Configuration in Pigeons and Humans: II. Generality across Search Tasks," *Journal of Comparative Psychology* 111, no. 1 (1997), 14–24.

8 Gladwin wrote of his extensive experiences among the Puluwat and what he learned about their means of navigation in a wonderfully accessible book entitled *East Is a Big Bird: Navigation and Logic on Puluwat Atoll* (Harvard University Press: Boston, 1970).

9 David Lewis, *We, the Navigators: The Ancient Art of Landfinding in the Pacific* (University of Hawaii Press: Honolulu, 1972), 90.

10 The studies of spatial memory in Australian Aboriginal children were carried out by Judith Kearins of the University of Western Australia and published in the report "Visual Spatial Memory in Australian Aboriginal Children of Desert Regions," *Cognitive Psychology* 13, no. 3 (1981), 434–460.

11 Chatwin's book *The Songlines* (Penguin: New York, 1988) is worth reading as both a narrative description of the Australian Aboriginal's connection to the land and also as an extended meditation on the relationship between human beings and the natural world.

CHAPTER 3: LOOKING FOR ROUTES

1 The description of Edward Atkinson's struggle can be found on page 303 of Apsley Cherry-Garrard's extraordinary memoir of the Scott expedition,

The Worst Journey in the World (Chatto & Windus: London, 1965). The book contains a wealth of details describing the influence of the stark and feature-less terrain of the Antarctic on human sensory perception.

2 The study of lost-person behavior in Peter Lougheed Park was conducted by David Heth of the University of Alberta and described in the paper "Char-acteristics of Travel by Persons Lost in Albertan Wilderness Areas," *Journal of Environmental Psychology* 18 (1998), 223–235.

3 Ronald Schmidt and Dwight McCarter, *Lost! A Ranger's Journal of Search and Rescue* (Graphicom: Yellow Springs, OH, 1988).

4 Many of Wehner's classic studies are described in his paper with Martin Muller entitled "Path Integration in Desert Ants, *Cataglyphis fortis,*" *Proceed-ings of the National Academy of Sciences* 85, no. 4 (1988), 5287–5290.

5 The studies showing that ants with backpacks still estimate distance accu-rately are described in a chapter called "Arthropods" written by Wehner for the book *Animal Homing*, edited by F. Papi (Chapman and Hall: London, 1992), 45–144. The role of optic flow in ant navigation is described in a paper by B. Ronacher and R. Wehner entitled "Desert Ants *Cataglyphis fortis* Use Self-induced Optic Flow to Measure Distances Travelled," *Journal of Comparative Physiology A* 177 (1995), 21–27.

6 Wehner's leg extension and amputation experiments are described in the paper by M. Wittlinger, R. Wehner, and H. Wolf entitled "The Ant Odometer: Stepping on Stilts and Stumps," Science 312, no. 5782 (2006), 1965–1967. The finding that ants can incorporate changing altitudes in their homing vectors can be found in G. Grah, R. Wehner, and B. Ronacher, "Path Inte-gration in a Three-dimensional Maze: Ground Distance Estimation Keeps Desert Ants *Cataglyphis fortis* on Course," *Journal of Experimental Biology* 208 (2005), 4005–4011.

7 The classic demonstration of path integration in nursing mother gerbils is described in M. L. Mittelstaedt and H. Mittelstaedt's "Homing by Path Inte-gration in a Mammal," *Naturwissenschaften* 67, no. 11 (1980), 566–567.

8 Path integration experiments in a variety of animals are described in the chapter by A. S. Etienne et al., "The Role of Dead Reckoning in Navigation," in *Spatial Representation in Animals*, edited by Susan Healy, (Oxford Univer-sity Press: Oxford, 1998), pp. 54–68.

9 A nice online biography of Robert Goddard can be found at www-istp.gsfc. nasa.gov/stargaze/Sgoddard.htm.

10 Experiments showing that visual fixes reset integration drift are described in A. S. Etienne, R. Maurer, and V. Seguinot's paper "Path Integration in Mammals

and Its Interaction with Visual Landmarks," *Journal of Experimental Biology* 199, no. 1 (1996), 201–209. and in A. S. Etienne et al., "A Brief View of Known Landmarks Reorientates Path Integration in Hamsters," *Naturwissenschaften* 87, no. 11 (2000), 494–498.

11 Ursula von St. Paul's studies of visual path integration in geese were originally reported in her "Do Geese Use Path Integration for Walking Home?" in *Avian Navigation*, edited by F. Papi and H. G. Wallraff (Springer: Berlin, 1982), 298–307.

12 The original blindwalking studies were conducted and described by J. Thomson in "Is Continuous Visual Monitoring Necessary in Visually Guided Locomotion?" *Journal of Experimental Psychology: Human Perception and Performance* 9 (1983), 427–443. Our own first foray into human path integration is described in M. G. Bigel and C. G. Ellard, "The Contribution of Nonvisual Information to Simple Place Navigation and Distance Estimation: An Examination of Path Integration," *Canadian Journal of Psychology* 54, no. 3 (2000), 172–185.

13 Experiments assessing humans' path-integration abilities can be found in J. M. Loomis et al., "Nonvisual Navigation by Blind and Sighted: Assessment of Path Integration Ability," *Journal of Experimental Psychology: General* 122 (1993), 73–91.

CHAPTER 4: MAPS IN THE WORLD

1 Readers looking for a more detailed treatment of some of the concepts of topology could do no better than to look up Jeffrey Weeks's *The Shape of Space* (CRC Press: Boca Raton, 2001). It is a *very* entertaining math book (I'm not kidding!) that takes the reader from basic concepts in topology to discussions of experiments and observations that may reveal the shape of the universe.

2 The ancient history of pigeons is described in J. Hermans, *The Handbook of Pigeon Racing* (Pelham Books: London, 1986).

3 An excellent review of navigation in birds including pigeons can be found in the online book *Animal Spatial Cognition: Comparative, Neural and Computational Approaches*, edited by Michael Brown and Robert Cook and presented by the Society for Comparative Cognition (www.pigeon.psy.tufts. edu/asc/toc.htm). The chapter by Verner Bingman entitled "Behavioral and Neural Mechanisms of Homing and Migration in Birds" (www.pigeon.psy. tufts.edu/asc/Bingman/Default.htm) provides a good overview. The chapter by John Phillips, Klaus Schmidt-Koenig, and Rachel Muheim entitled

"True Navigation: Sensory Bases of Gradient Maps" (www.pigeon.psy.tufts. edu/asc/Phillips/Default.htm) provides a thorough explanation of gradient map concepts.

4 The study of pigeon homing at the Auckland Junction Magnetic Anomaly can be found in T. E. Dennis, M. J. Rayner, and M. M. Walker, "Evidence That Pigeons Orient to Geomagnetic Intensity during Homing," *Proceedings of the Royal Society B* 274, no. 1614 (2007), 1153–1158.

5 Hans Wallraff reviews his odor theory of pigeon navigation in a paper entitled "Avian Olfactory Navigation: Its Empirical Foundation and Conceptual State," *Animal Behaviour* 67, no. 2 (2004), 189–204.

6 Carl Cornelius's sea turtle story was originally recorded in his work *Die Zug- und Wander-Thiere aller Thierclassen* (Springer: Berlin, 1865) but came to me secondhand in the paper by L. Avens and K. Lohmann entitled "Navigation and Seasonal Migratory Orientation in Juvenile Sea Turtles," *Journal of Experimental Biology* 207 (2004), 1771–1778.

7 Some aspects of turtle navigation are described in papers such as K. Lohmann et al., "Animal Behaviour: Geomagnetic Map Used in Sea-turtle Navigation," Nature 428, no. 6986 (2004), 909–910, and K. Lohmann and C. Lohmann, "Orientation and Open-Sea Navigation in Sea Turtles," *Journal of Experimental Biology* 199, part 1 (1996), 73–81.

8 Robin Baker's studies of human magnetic sense are described in his book *Human Navigation and the Sixth Sense* (Simon and Schuster: New York, 1981).

9 The finding of magnetic bones in the human head was reported by R. Baker, J. G. Mather, and J. H. Kennaugh in a paper entitled "Magnetic Bones in Human Sinuses," *Nature,* 301 (1983), 78–80.

10 Finney describes his work with Nainoa Thompson and the Polynesian Voyaging Society in *Voyage of Rediscovery: A Cultural Odyssey through Polynesia* (University of California Press: Berkeley, 1994). Finney speculates that Thompson may at times have been guided by magnetoreception in his paper "A Role for Magnetoreception in Human Navigation?" *Current Anthropology* 36 (1995), 500–506.

11 David Lewis's story of scrotal navigation is related in his book *We, the Navigators: The Ancient Art of Landfinding in the Pacific* (University Press of Hawaii: Honolulu, 1972), 86–92.

12 Descriptions of Bedouin tracking methods are found in Donald P. Cole, *Nomads of the Nomads: The Al Murrah Bedouin of the Empty Quarter* (Aldine Publishing: Chicago, 1975).

13 Wilfred Thesiger, *Arabian Sands* (Dutton: New York, 1959), 51–52.

CHAPTER 5: MAPS IN MOUSE MINDS

1 The best biography of Edward Chace Tolman is found in volume 1 of *Portraits of Pioneers in Psychology*, edited by G. A. Kimble, M. Wertheimer and C. L. White (APA Press: Washington, DC, 1991) in a chapter by Henry Gleitman entitled "Edward Chace Tolman: A Life of Scientific and Social Purpose," 227–241.

2 Tolman's classic starburst maze experiments are described in his paper "Cognitive Maps in Rats and Men," *Psychological Review 55* (1948), 189–208.

3 The story of von Frisch's marvelous early experiments with bees is told by him in the wonderful little book *Bees: Their Vision, Chemical Senses, and Language* (Cornell University Press: Ithaca, NY, 1972).

4 A brief discussion of Aristotle's observations of bees is given by the biologist J. B. S. Haldane in "Aristotle's Account of Bees' 'Dances,'" *The Journal of Hellenic Studies 75* (1955), 24–25.

5 A nice illustrated account of Michelsen's robot bee studies is given in the article by J. Knight entitled "Animal Behaviour: When Robots Go Wild," *Nature 434*, no. 7036 (2005), 954–955.

6 J. L. Gould, "Honey Bee Cognition," *Cognition 37* (1990), 83–103.

7 The paper by R. Menzel et al. entitled "The Knowledge Base of Bee Navigation," *Journal of Experimental Biology 199*, no. 1 (1996), 141–146, gives a good overview of some recent thinking about bee maps.

8 It took a large team to demonstrate, using sophisticated radar tracking methods, that bees might, after all, have a "map-like" spatial memory. See R. Menzel et al., "Honey Bees Navigate According to a Map-like Spatial Memory," *Proceedings of the National Academy of Sciences 102*, no. 8, (2005), 3040–3045.

9 David Sherry has written a nice, accessible review of food caching in birds: "Food Storing in the Paridae," *Wilson Bulletin 101*, no. 2 (1989), 289–304.

10 The studies on spatial mapping in Clark's nutcrackers were conducted by Brett Gibson and Alan Kamil and described in "Tests for Cognitive Mapping in Clark's Nutcrackers (*Nucifraga columbiana*)," *Journal of Comparative Psychology 115*, no. 4 (2001), 403–417.

CHAPTER 6: MUDDLED MAPS IN HUMAN MINDS

1 Jean Piaget and Barbel Inhelder wrote the classic first book on the development of spatial concepts in children, *The Child's Conception of Space*, translated by F. J. Langdon and J. L. Lunzer (Routledge and Kegan Paul: London, 1971).

2 For the history of maps, I can do no better than recommend the massive volumes of the History of Cartography Project. The material I discuss in this

chapter was taken from volume 1 of *Cartography in Prehistoric, Ancient, and Medieval Europe and the Mediterranean,* edited by J. Brian Harley and David Woodward (University of Chicago Press: Chicago, 1987).

3 It is easy to find sensational and preposterous accounts of the origin and purpose of the Nazca lines. A more balanced treatment, which embeds the lines in their full cultural context, can be found in Helaine Silverman and Donald Proulx's academic but quite readable book *The Nasca* (Wiley-Blackwell: Hoboken, NJ, 2002).

4 Barbara Tversky has contributed an enormous number of insightful articles and presentations to our understanding of human spatial cognition, among many other things. One of my favorites, which contains most of the ideas discussed here, is the chapter "Cognitive Maps, Cognitive Collages, and Spatial Mental Models," in *Spatial Information Theory: A Theoretical Basis for GIS,* edited by A. U. Frank and I. Campari (Springer-Verlag: Berlin, 1995), 14–24.

5 Erik Jonsson's book Inner Navigation: *Why We Get Lost and How We Find Our Way* (Scribner: New York, 2002) is a delightful general account of some aspects of human navigation. Jonsson is very much of the belief that experiments on urban university students are preventing us from understanding the true navigational capability of human beings. He even has a hunch that Robin Baker was on the right track with his magnetoreception studies and has encouraged me to continue with them. I'd almost like to do so just for the fun of watching participants stumble through forests with magnets on their heads!

6 Yi-fu Tuan's *Space and Place: The Perspective of Experience* (University of Minnesota Press: Minneapolis, 2001) is a remarkable work of anthropology covering much of the relationship between the basic dimensions of space and our beliefs, feelings, thoughts, and rituals.

7 Some of the experiments dealing with accessing spatial information between nested environments can be found in the paper by F. Wang and J. Brockmole entitled "Switching between Environmental Representations in Memory," *Cognition* 83 (2002), 295–316.

8 Tim McNamara reported the work on the structure of spatial memories for objects presented on a screen in a paper with J. K. Hardy and S. C. Hirtle entitled "Subjective Hierarchies in Spatial Memory," *Journal of Experimental Psychology: Learning, Memory and Cognition* 15, no. 2 (1989), 211–227.

9 Jean Piaget and Barbel Inhelder, *The Child's Conception of Space,* translated by F. J. Langdon and J. L. Lunzer (Routledge and Kegan Paul: London, 1971).

CHAPTER 7: HOUSE SPACE

1 Research on the effectiveness of home staging is done (not surprisingly) by realtors and home stagers. One oft cited finding, based on a 2003 U.S. survey by the real estate brokerage company HomeGain, suggests that home staging produced an average return on investment of 169 percent, where the price range of staging in the survey ranged from about $200 to $1,000. Other anecdotal accounts suggest that staging also shortens the average time that a house is on the market.

2 Isovists in architecture were first described in a paper by M. L. Benedikt entitled "To Take Hold of Space: Isovists and Isovist Fields," *Environment and Planning B, Planning and Design* 6, no. 1 (1979), 47–65.

3 Some of the studies showing that isovist shapes influence feeling and behavior are described in a report by G. Franz and J. M. Wiener entitled "Exploring Isovist-based Correlates of Spatial Behavior and Experience," in *Proceedings of the 5th International Space Syntax Symposium* (2005), 503–517.

4 Jay Appleton's arguments about the importance of prospect and refuge to the human psyche are laid out in his book *The Experience of Landscape* (Wiley: Hoboken, NJ, 1975).

5 Winnifred Gallagher's *House Thinking*, a delightful tour through the main parts of a modern house, is filled with much interesting material on the history and philosophy of house building (Harper Perennial: New York, 2007).

6 Christopher Alexander, in *A Pattern Language: Towns, Building, Constructions* (Oxford University Press: Oxford, 1977), lays out an encyclopedic set of intuitive rules governing how welcoming and functional space should be organized. The theory underlying the rules is described in several volumes, including *The Timeless Way of Building* (Oxford University Press: Oxford, 1979).

7 Amos Rapoport's seminal work on the meaning of house form, written from a cross-cultural perspective, is his book *House Form and Culture* (Prentice-Hall: New York, 1969).

8 It is easy to find bad information about feng shui and difficult to find useful information. Two books that I found useful were Cate Bramble's *The Architect's Guide to Feng Shui: Exploding the Myth* (Architectural Press: Oxford, 2003) and Kartar Diamond's *Feng Shui for Skeptics: Real Solutions without Superstition* (Four Pillars Publishing: Los Angeles, 2003).

9 Christopher Alexander, Notes on the Synth*esis of Form* (Harvard University Press: Boston, 1964).

10 Alexander's *The Nature of Order* (volume 1, *The Phenomenon of Life*; volume 2, *The Process of Creating Life*; volume 3, *A Vision of the Living World*; and volume 4, *The Luminous Ground*) was published in 2004 by the Center for Environmental Structure, Berkeley, California.

11 Sarah Susanka, *The Not So Big House: A Blueprint for the Way We Really Live* (Taunton Press: Newtown, CT, 2001).

12 Muthesius's magnum opus, *Das Englisch Haus*, was originally published in 1904. An English translation by Janet Seligman and Stewart Spencer, *The English House*, has recently been published in a beautiful edition with all his original drawings intact (Frances Lincoln: London, 2006).

13 Muthesius, *Layout and Construction*, volume 2 of *The English House*, p. 9.

CHAPTER 8: WORKING SPACE

1 A nice summary of Propst's Action Office concept and his attitude toward its current "Dilbert-ization" can be found in the 1998 article by Yvonne Abraham in *Metropolis Magazine* entitled "The Man behind the Cubicle," www.metropolismag.com/html/content_1198/n098man.htm.

2 A discussion of visibility graphs can be found in the paper by A. Turner et al. entitled "From Isovists to Visibility Graphs: A Methodology for the Analysis of Architectural Space," *Environment and Planning B: Planning and Design* 28 (2001), 103–121.

3 Though his language is complex at times, Bill Hillier's books on space syntax were the texts that began the space syntax movement. The key books are *The Social Logic of Space*, written with J. Hanson (Cambridge University Press: Cambridge, 1984) and *Space Is the Machine: A Configurational Theory of Architecture* (Cambridge University Press: Cambridge, 1996). The main fountain of information about space syntax can be found at the website of the original group of researchers at the Bartlett School of Planning (www.spacesyntax.org).

4 One technical paper describing the use of computer-based autonomous agents to model human behavior in space is A. Turner and A. Penn's "Encoding Natural Movement as an Agent-based System: An Investigation into Human Pedestrian Behaviour in the Built Environment," *Environment and Planning B: Planning and Design* 29, no. 4 (2002), 473–490.

5 Some of the tricks of the trade used to engineer space to influence consumer behavior are described in Paco Underhill's book *The Call of the Mall* (Simon and Schuster: New York, 2004).

6 A discussion of social control in food courts can be found in the article by John Manzo entitled "Social Control and the Management of 'Personal'

Space in Shopping Malls," *Space and Culture* 8 (2005), 83–97.

7 Bill Friedman's bible of casino design, *Designing Casinos to Dominate the Competition* (Institute for the Study of Gambling and Commercial Gaming: Las Vegas, 2000), consists of a few introductory chapters that outline his rules of design, followed by extensive case studies of Las Vegas casinos.

8 Kranes's playground approach to casino design is described in his paper "Playgrounds," published in the *Journal of Gambling Studies* 11 (1995), 91–102. Kranes describes casinos as "managed wildness" in an online essay entitled "Toward More Adventurous Playgrounds: Casino Lost; Casino Regained," www.unr.edu/gaming/papers/kranes.asp.

9 The work of Finlay and her team is described in two papers: K. Finlay et al., "The Physical and Psychological Measurement of Gambling Environments," *Environment and Behavior* 38, no. 4 (2006), 570–581, and K. Finlay *et al.,* "Trait and State Emotion Congruence in Simulated Casinos," *Journal of Environmental Psychology* 27 (2007), 166–175.

10 Recent developments and reactions to cubicle hive office designs are described in an article written by Julie Schlosser in *Fortune Magazine*'s online portal, CNNMoney.com, entitled "Cubicles: The Great Mistake" (March 22, 2006), http://money.cnn.com/2006/03/09/magazines/fortune/cubicle_howi-work_fortune/index.htm.

11 The Chiat-Day experience is documented in a paper by W. R. Sims, M. Joroff, and F. Becker, "Teamspace Strategies: Creating and Managing Environments to Support High Performance Teamwork," IDRC Foundation, Atlanta, 1998.

12 The idea of the quotation and Heerwagen's "cognitive cocoon" come from J. H. Heerwagen et al., "Collaborative Knowledge Work Environments," *Building Research and Information* 32, no. 6 (2004), 510–528.

13 The studies of the effects of physical proximity on communication in the two research laboratories are described by R. E. Kraut, C. Egido, and J. Galegher, "Patterns of Contact and Communication in Scientific Research Collaboration," in *Intellectual Teamwork*, edited by Galegher, Kraut, and Egido (L. Erlbaum: Hillsdale, NJ, 1990).

14 Early classic studies of the effects of proximity are described in Thomas Allen's paper "Communications Networks in R&D Laboratories," *R&D Management* 1, no. 1 (1970), 14–21.

15 The case study of the redesign of the ThoughtForm offices is found in the paper by J. Peponis *et al.,* "Designing Space to Support Knowledge Work," *Environment and Behavior* 39 (2007), 815–841.

CHAPTER 9: CITY SPACE

1 Nico Oved's description of his photographic exposition of housing in *les banlieues* is described at his website, www.nicooved.com.

2 Although Le Corbusier's urban planning principles caused much damage both overseas and at home (largely because of his failure to understand the psychology of space), he is still widely respected as an artist and architect. A sympathetic and interesting book is W. Boesiger and H. Girsberger's *Le Corbusier 1910–65* (Birkhäuser: Basel, 1999), which contains photographs of his works along with captions in three languages explaining his intentions.

3 Jane Jacobs condemns Le Corbusier's mathematics, and some other aspects of his urban-planning principles, in the introductory chapter of her opus *The Death and Life of Great American Cities*, revised edition (Vintage: New York, 1992).

4 The quote from Oscar Newman comes from page 10 of his book *Creating Defensible Space* (Center for Urban Policy Research, U.S. Department of Housing and Urban Development: Washington D.C., 1996).

5 Christian Nold's exploits can be found on his website, www.biomapping.net.

6 The material on Fetter Lane is found in chapter 22, "A London Address," of Peter Ackroyd's *London: The Biography* (Vintage, 2001). The quote is taken from page 2.

7 The biographical details on Ivan Chtcheglov and Guy Debord are taken from Merlin Coverley's book Psychogeography (Pocket Essentials: London, 2007). The quotes from Chtcheglov come from the Situationist *International Anthology*, edited by Ken Knabb (Bureau of Public Secrets: Berkeley, 1981), 1–2.

8 Kevin Lynch, *The Image of the City* (MIT Press: Cambridge, MA, 1960).

9 Biographical details of William Whyte, and a description of his novel methods applied to understanding public space, can be found in *The Essential William H. Whyte,* edited by Albert LaFarge (Fordham University Press: New York, 2000). The most interesting material from the standpoint of this chapter is the excerpt from Whyte's book *The Social Life of Small Urban Spaces* (Conservation Foundation, 1980; reprinted by the Project for Public Spaces: New York, 2001).

10 Raymond Curran's *Architecture and the Urban Experience* (Van Nostrand Reinhold: New York, 1983) was a lucky find for me that came about when his niece, a good friend of mine, heard about the subject of my book and thrust her uncle's work into my hands.

11 Jan Gehl, *Life between Buildings: Using Public Space* (Van Nostrand Reinhold: New York, 1987).

12 I mentioned Bill Hillier's books (*The Social Logic of Space* and *Space Is the Machine*) in the previous chapter. These important works also support what I have said here.

13 The map of my neighborhood was produced using the software webmap-AtHome, written by N. S. C. Dalton of the University College London Virtual Reality Centre for the Built Environment.

14 A useful paper describing the relationship between space syntax and spatial cognition is by Alan Penn, "Space Syntax and Spatial Cognition," *Environment and Behavior* 35 (2003), 30–65.

15 An interesting set of studies showing the influence of the shape of pedestrian spaces on the perception of space and time can be found in the article by Raymond Isaacs entitled "The Subjective Duration of Time in the Experience of Urban Places," *Journal of Urban Design* 6, no. 2 (2001), 109–127.

16 The idea that we learn new spaces by working outwards from a skeleton of well-integrated areas was first described by B. Kuipers, D. Tecuci, and B. Stankiewicz in the paper "The Skeleton in the Cognitive Map," *Environment and Behavior* 35, no. 1 (2003), 81–106.

17 Hillier's deformed wheel theory of the city form is found in his article "A Theory of the City as Object," published in *Proceedings of the 3rd International Space Syntax Symposium 2001*. The quote is found on page 2.27 of that article. The maps of London and Tokyo are taken from an article by Hillier and Laura Vaughan entitled "The City as One Thing," published in *Progress in Planning* 67, no. 3 (2007), 205–230, also available at http://eprints.ucl. ac.uk/archive/00003272/01/3272.pdf.

18 The interesting space syntax of the city of Nicosia is described by Bill Hillier and Laura Vaughan in the article "The City as One Thing," *Progress in Planning*, 67, no. 3 (2007), 205–230.

19 An excellent summary of the London congestion pricing experience, its effects, and implications for similar schemes in other cities has been published by Todd Litman of the Victoria Transport Policy Institute of Canada, www.vtpi.org/london.pdf. An analysis of London's experience and its implications for the proposal of a similar scheme for Manhattan is reported by the New York Academy of Sciences in an e-briefing written by Christine Van Lenten entitled "Congestion Pricing for New York? Lessons from London" (May 10, 2007), available at www.nyas.org/ebrief/miniEB.asp?eBriefID=644.

20 The controversy surrounding congestion pricing in Manhattan is described in the *New York Times* article by Diane Cardwell entitled "Faster, Maybe. Cheaper, No. But Driving Has Its Fans" (March 31, 2008).

The Partnership for New York City survey results are available at www. pfnyc.org/pressReleases/2007/PFNYC%20Driver%20Survey%20Results.pdf.

21 Jane Jacobs describes her run-in with traffic engineers in her book *Dark Age Ahead* (Vintage: New York, 2005).

22 I have relied heavily on Howard Kunstler's brilliant and influential *The Geography of Nowhere: The Rise and Decline of America's Man-Made Landscape* (Free Press: New York, 1994).

23 A fascinating repository of facts and figures related to the worldwide problem of urban sprawl can be found in the book by Jeffrey Kenworthy and Felix Laube entitled *An International Sourcebook of Automobile Dependence in Cities, 1960–1990* (University Press of Colorado: Boulder, CO, 1999).

24 One of the most widely cited (and frightening) views of the coming changes related to peak oil is Howard Kunstler's *The Long Emergency: Surviving the End of Oil, Climate Change, and Other Converging Catastrophes of the Twenty-first Century* (Grove Press: New York, 2006).

25 The history of Portland's experiences with Oregon's groundbreaking 1973 restrictions on urban sprawl, as well as many other positive examples of smart growth approaches, can be found in the book by F. Kaid Benfield et al., *Solving Sprawl: Models of Smart Growth in Communities Across America* (Island Press: Washington, DC, 2001).

26 The Ontario Places to Grow Act is described comprehensively on the Government of Ontario's website. The bill itself can be found at www.ontla. on.ca/bills/bills-files/38_Parliament/Session1/b136ra.pdf.

CHAPTER 10: CYBERSPACE

1 Interview with Philip Rosedale conducted by Michael Fitzgerald entitled "How I Did It," *Inc. Magazine*, February 2007.

2 Edward Hall's groundbreaking studies on proxemics are described in his *The Hidden Dimension* (Anchor Books: New York, 1966).

3 The study on proxemics in Second Life was published by Nick Yee et al. as "Unbearable Likeness of Being Digital: The Persistence of Nonverbal Social Norms in Online Virtual Environments," *CyberPsychology & Behavior* 10, no. 1 (2007), 115–121.

4 Joshua Meyrowitz, *No Sense of Place: The Impact of Electronic Media on Social Behavior* (Oxford University Press: Oxford, 1986).

5 The U.S. government website www.gps.gov contains much useful background information on the GPS signal and how it can be used. A clear textbook approach can be found in Ahmed El-Rabbany's *Introduction to GPS:*

The Global Positioning System, 2nd edition (Artech: Norwood, MA, 2006).

6 As I write these words, a news story has hit the airwaves describing how an American student who was jailed in Egypt was able to alert his network of friends by sending a one-word message—"arrested"—to Twitter using his cell phone. Like the efforts of Tibetan protesters to subvert attempts to censor the Internet, this story shows how cyberspace can change politics by collapsing real space and time. An insightful article describing this episode and some other real-world uses of Twitter was written by Murray Whyte in an article entitled "Tweet Tweet—There's Been an Earthquake" *Toronto Star*, June 1, 2008.

7 A succinct summary of ubiquitous computing approaches can be found in Mark Weiser and John Seely Brown, "Designing Calm Technology," at www.ubiq.com/weiser/calmtech/calmtech.htm.

8 The Ambient Orb is described at www.ambientdevices.com/cat/products.html.

9 Mark Weiser and John Seely Brown, "The coming age of calm technology," in Peter J. Denning and Robert M. Metcalfe's *Beyond Calculation: The Next 50 Years of Computing* (Springer: New York, 1998), p.81.

10 The website of Waterloo's Research Laboratory for Immersive Virtual Environments is http://virtualpsych.uwaterloo.ca.

11 The virtual wormhole study is described in B. Schnapp and W. Warren, "Wormholes in Virtual Reality: What Spatial Knowledge Is Learned for Navigation?" abstract, Journal of Vision 7, no. 9 (2007), 758, 758a; http://journalofvision.org/7/9/758.

12 Some of Philip Beesley's work is described in the beautiful book edited by Beesley et al. entitled *Responsive Architectures: Subtle Technologies* (Riverside Architectural Press: Riverside, CA, 2006).

13 Blascovich personal communication, December 15, 2006.

14 Many details and photos of Heilig's Sensorama can be found at www.mortonheilig.com.

15 One description of racist hate in digital Darfur is given in the Second Life–related blog New World Notes, in an entry from May 1, 2006, by Wagner James Au entitled "Guarding Darfur" (http://nwn.blogs.com/nwn/2006/05/guarding_darfur.html).

16 Paul Virilio's apocalyptic vision of the influence of technology and speed on life is given in his Open Sky, translated by Julie Rose (Verso: London, 1997). His opinions on the influence of telecommunication technologies on the conduct of warfare are found in *Desert Screen: War at the Speed of Light*, translated by Michael Degener (Continuum: London, 2005).

CHAPTER 11: GREENSPACE

1 Bruce Chatwin tells the story of accelerated songlines in his book *The Songlines* (New York: Penguin, 1988).

2 Jane Jacobs describes the origins of modern suburban living in the introductory chapter of her book *The Death and Life of Great American Cities* (Vintage: New York, 1992).

3 Mathis Wackernagel and William Rees, *Our Ecological Footprint: Reducing Human Impact on the Earth* (New Society Publishers: Gabriola Island, BC, 2001).

4 Details of Wilson's life can be found in his book *Naturalist* (Shearwater Books: Washington, DC, 1996).

5 Wilson's book *Biophilia* (Harvard University Press: Cambridge, MA, 1986) started things off, but much useful followup material can be found in the later book edited by Stephen Kellert and Edward Wilson and entitled *The Biophilia Hypothesis* (Island Press: Washington, DC, 1995).

6 Many studies describing the human preference for natural settings can be found in Stephen Kaplan's book *The Experience of Nature: A Psychological Perspective* (Cambridge University Press: New York, 1989).

7 Many of the findings relating to the beneficial effects of exposure to nature on healing, productivity, and happiness are reviewed in Stephen Kellert and Edward Wilson's edited volume, *The Biophilia Hypothesis* (Island Press: Washington, DC, 1995).

8 Richard Louv, *Last Child in the Woods: Saving Our Children from Nature-Deficit Disorder* (Algonquin Books: New York, 2006).

9 Erik Jonsson, *Inner Navigation: Why We Get Lost and How We Find Our Way* (Scribner: New York, 2002).

10 Tim Lougheed, "The Not-So-Great Outdoors," *University Affairs*, April 2006.

11 Information on the engaging pursuit of geocaching can be found at www.geocaching.com.

12 Stephen Kellert's biophilic design book *Building for Life: Designing and Understanding the Human–Nature Connection* (Island Press: Washington, DC, 2005) contains much interesting material on how to construct buildings and towns to encourage connections with nature.

13 The Meadows, and many other kid-friendly biophilic neighborhood designs, are described in Robin Moore and Clare Cooper Marcus's chapter "Healthy Planet, Healthy Children: Designing Nature into the Daily Spaces of Childhood," in *Biophilic Design: The Theory, Science, and Practice of Bringing Buildings to Life*, edited by S. Kellert, J. H. Heerwagen, and M. L. Mador (Wiley: Hoboken, NJ, 2008), 153–204.

14 Barry Blesser and Linda-Ruth Salter, *Spaces Speak, Are You Listening?: Experiencing Aural Architecture* (MIT Press: Cambridge, MA, 2006). Some ideas about how to engage senses other than the visual in architectural design are contained in architect Juhani Pallasmaa's seminal book *The Eyes of the Skin: Architecture and the Senses* (Academy Editions: London, 2005).

15 The Conflux Festival's website is at www.confluxfestival.org. Polli's *NYSoundmap* is described at www.nysoundmap.org.

16 Stephen Jay Gould, *Eight Little Piggies: Reflections in Natural History* (Norton: New York, 1993).

CHAPTER 12: THE FUTURE OF SPACE

1 Michael Jones describes the philosophy and impact of Google Earth in his "Google's Geospatial Organizing Principle," *IEEE Computer Graphics and Applications* 27, no. 4 (2007), 8–13.

2 The Crisis in Darfur project can be found at www.ushmm.org/googleearth/projects/darfur/.

3 An interesting book describing the recent history and use of mazes is *Walking a Sacred Path: Rediscovering the Labyrinth as a Spiritual Tool*, by Lauren Artress (Riverhead Books: New York, 1995).

4 Charlene Spretnak describes the relationship between the Chiapas revolution and the forces of globalization (along with several other interesting movements suggestive of a new appreciation for the importance of place) in her book *The Resurgence of the Real: Body, Nature and Place in a Hypermodern World* (HarperCollins Canada: Toronto, 1998).

5 Alisa Smith leads the way with her *100 Mile Diet: A Year of Local Eating* (Random House: New York, 2007). Another initiative with similar motivations is Sarah Bongiorni's *A Year without "Made in China": One Family's True Life Adventure in the Global Economy* (Wiley: Hoboken, NJ, 2007).

6 Peter Mayle, *A Year in Provence* (Vintage: New York, 1991).

INDEX